DC Circuit Basics

FIRST EDITION

BY PRASUN BARUA

ABOUT

Welcome to DC Circuit Basics! This is a nonfiction science book which contains various topics on basics of DC circuit. Direct Current (D.C) is a form of electrical current which flows around an electrical circuit in one direction only, making it a "Uni-directional" supply. It does not regularly change direction. This book contains various topics like Theory Of DC Circuit, Ohm's Law And Power, Electrical Units Of Measure, Kirchhoff's Circuit Law, Mesh Current Analysis, Nodal Voltage Analysis, Thevenin's Theorem, Norton's Theorem, Maximum Power Transfer, Star Delta Transformation, Voltage Sources, Current Sources, Kirchhoff's Current Law, Kirchhoff's Voltage Law, Voltage Dividers, Current Dividers, Electrical Energy And Power, DC Circuit And Waveform. This is the first edition of the book. Thanks for reading the book.

TABLE OF CONTENTS

CHAPTER NO.	TITLE	PAGE NO.
CHAPTER-1	THEORY OF DC CIRCUIT	4
CHAPTER-2	OHM'S LAW AND POWER	12
CHAPTER-3	ELECTRICAL UNITS OF MEASURE	18
CHAPTER-4	KIRCHHOFF'S CIRCUIT LAW	19
CHAPTER-5	MESH CURRENT ANALYSIS	25
CHAPTER-6	NODAL VOLTAGE ANALYSIS	31
CHAPTER-7	THEVENIN'S THEOREM	34
CHAPTER-8	NORTON'S THEOREM	39
CHAPTER-9	MAXIMUM POWER TRANSFER	44
CHAPTER-10	STAR DELTA TRANSFORMATION	51
CHAPTER-11	VOLTAGE SOURCES	61
CHAPTER-12	CURRENT SOURCES	76
CHAPTER-13	KIRCHHOFF'S CURRENT LAW	89
CHAPTER-14	KIRCHHOFF'S VOLTAGE LAW	100
CHAPTER-15	VOLTAGE DIVIDERS	106
CHAPTER-16	CURRENT DIVIDERS	124
CHAPTER-17	ELECTRICAL ENERGY AND POWER	137
CHAPTER-18	DC CIRCUIT AND WAVEFORM	144

CHAPTER-1: THEORY OF DC CIRCUIT

VOLTAGE, CURRENT & RESISTANCE

In an electrical or electronic circuit, Ohm's Law describes the fundamental relationship between voltage, current, and resistance. Atoms make up all materials, and protons, neutrons, and electrons are found in every atom. The electric charge of protons is positive. Electrons have a negative electrical charge, but Neutrons have none. Atoms are held together by strong magnetic forces between the nucleus and the electrons in the outer shell.

The atom is steady when these protons, neutrons, and electrons are all composed. However, if we separate them from one another, they will rejoin and begin to exert a potential attraction known as a potential difference.

Due to their attraction, these loose electrons will start to travel and drift back to the protons if we establish a closed circuit, resulting in an electron flow. An electrical current is the flow of electrons. The electrons cannot easily travel across the circuit because the material they pass through restricts their flow. Resistance is the term for this constraint. There are three different but closely associated electrical quantities in all basic electrical or electronic circuits: voltage (V), current (I) and resistance (Ω).

Voltage

The potential energy of a power supply kept in the form of an electrical charge is known as voltage (V). The force that moves electrons through a conductor is known as voltage, and the higher the voltage, the greater the ability to "move" electrons through a circuit.

Potential energy can be defined as the amount of work required in joules to move electrons in the form of an electrical current around a circuit from one point or node to another. The Potential Difference, (p.d.) commonly referred to as the Voltage Drop, is the difference in voltage between any two points, connections, or junctions (referred to as nodes) in a circuit.

Though "energy (E)" is occasionally used to signify a generated electromotive force (emf), the potential difference between two points is measured in volts, with the circuit symbol "V".

The higher the voltage, the higher the pressure (or driving force) and the higher the work capacity. The term "DC Voltage" refers to a voltage source that is constant and unchanged across time.

An AC voltage is a voltage source with a periodic amplitude change over time. Voltage is measured in volts, which is defined as the electrical pressure required to drive an electrical current of one ampere through a resistance of one Ohm, regardless of whether the supply is AC or DC.

Voltages are commonly represented in Volts, prefixes like microvolts (V = 10^{-6} V), millivolts (mV = 10^{-3} V), and kilovolts (kV = 10^3 V) are used to specify sub-multiples of the voltage present. It's important to note that the amplitude of voltage might be either positive or negative.

A D.C. (direct current) voltage source with a fixed value and polarity is created by batteries, power sources, or solar cells. For instance, 5v, 12v, -9v, and so on. On the other hand, A.C. (alternating current) voltage sources used in homes, offices, and industrial settings, have a value that corresponds to the amount of power they supply.

In the United States, the voltage and frequency of mains alternating current (AC) electricity used in residences is commonly 110 volts AC (110V) and 230 volts AC (230V) in the United Kingdom.

Low voltage DC battery supplies of 1.5V to 24V dc are used in most electronic circuits. A constant voltage source's circuit symbol is commonly a battery icon with a positive, + and negative, − sign denoting polarity direction. An alternating voltage source is represented as a circle with a sine wave inside.

Symbols of Voltage

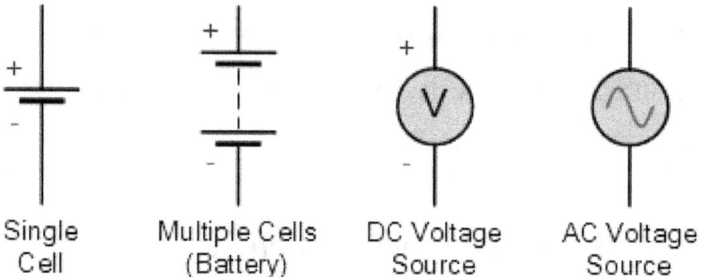

Single Cell   Multiple Cells (Battery)   DC Voltage Source   AC Voltage Source

An analogy can be established between a water tank and a power supply. The greater the pressure of the water as more energy is released, the higher the water tank above the outlet; the higher the voltage, the greater the potential energy as more electrons are released.

The voltage difference between any two places in a circuit is constantly measured, and the voltage difference between these two points is referred to as the "Voltage drop." It's important to remember that while voltage can exist across a circuit without current, current cannot exist without voltage. Therefore, open or semi-open circuit is preferable in any voltage source, whether DC or AC.

Current

The movement or flow of electrical charge is called current (I) which is measured in the units of Amperes (A). 1 ampere is 1 coulomb of electric charge per second travel through a point in a circuit.

The voltage source drives electrons (the negative particles of an atom) around a circuit in a continuous and uniform flow (known as a drift). In reality, electrons flow from the supply's negative (–ve) terminal to the positive (+ve) terminal, while traditional current flow implies that current flows from the positive to the negative terminal for ease of circuit understanding. Usually, an arrow having symbol of current (I) directs the actual path of the conventional current flow in any circuit diagram.

Conventional Current Flow

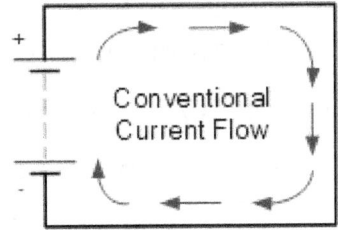

This is the movement of positive charge around a circuit from positive to negative. The diagram on the left depicts the passage of positive charge (holes) across a closed circuit, beginning at the positive terminal of the battery and ending at the negative terminal of the battery. Conventional current flow refers to the flow of current from positive to negative. This was the convention chosen during the discovery of electricity to describe the direction of electric current in a circuit.

To continue along this line of thought, the arrows depicted on symbols for components like as diodes and transistors in all circuit diagrams and schematics point in the direction of typical current flow. The flow of electrical current from positive to negative in the opposite direction of electron's actual flow is known as Conventional Current Flow.

Electron Flow

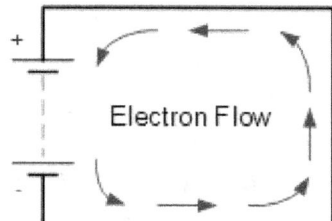

Electrons flow around the circuit in the reverse direction of typical current flow, which is negative to positive. The actual current flowing in an electrical circuit is comprised of electrons that travel from the negative pole of the battery (the cathode) to the positive pole (the anode). Since an electron's charge is negative by definition, it is drawn to the positive terminal. Electron Current Flow is the name given to this flow of electrons. As a result, electrons move from the negative terminal to the positive terminal of a circuit.

The direction of current flow has no effect on what happens within the circuit.

In general, understanding the conventional current flow — positive to negative — is much easier. A current source is a circuit device in electronic circuits that provides a specific quantity of current. For example, 1A, 5A, or 10A, with the circuit symbol for a constant current source shown as a circle with an arrow denoting its direction inside. An amp or ampere is defined as the number of electrons or charge (Q in Coulombs) traversing a specific location in the circuit in one second (t in Seconds).

Current is usually stated in Amps. In the case of micro amps it is expressed as $\mu A = 10^{-6}A$ and for milliamps it is $mA = 10^{-3}A$. Electrical current can be either positive or negative depending on the direction in which it flows around the circuit.

Direct Current, or D.C., is current that runs in one direction and Alternating Current, or A.C., is current that goes back and forth through the circuit. Whether AC or DC current flows through a circuit only when it is connected to a voltage source, with its "flow" limited by both the circuit's resistance and the voltage source driving it.

Furthermore, since alternating currents (and voltages) are periodic and change over time, the "effective" or "RMS," (Root Mean Squared) value expressed as I_{rms} produces the same average power loss as a DC current $I_{average}$. Current sources are the polar opposite of voltage sources in that they prefer short or closed circuits but despise open circuits since no current will flow.

Current is the analogue of the flow of water through the pipe using the tank of water relationship, with the flow being constant throughout the pipe. The greater the current, the faster the water flows. Since current cannot exist without voltage, every current source, whether DC or AC, prefers a short or semi-short circuit.

Resistance

When an element is capable to prevent or resist or prevent the flow of electric charge or current within a circuit is called Resistance (R). This circuit element is known the "Resistor". Resistance is measured in Ohms (Ω) with positive value only. 1 Kilo-ohms (kΩ) = $10^3 Ω$ and 1 Mega-ohms (MΩ) = $10^6 Ω$.

Symbols of Resistor

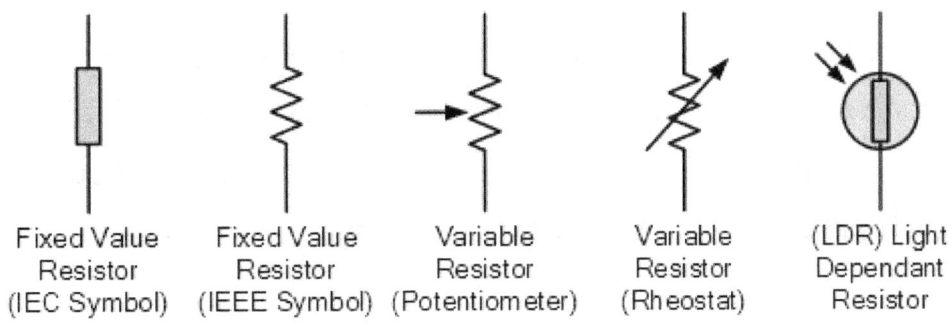

Fixed Value Resistor (IEC Symbol)   Fixed Value Resistor (IEEE Symbol)   Variable Resistor (Potentiometer)   Variable Resistor (Rheostat)   (LDR) Light Dependant Resistor

The correlation between the current flowing through a resistor and the voltage across it indicates whether the circuit element is a "good conductor" with low resistance or a "poor conductor" with high resistance.

Low resistance, such as 1 or less, indicates a good conductor made of materials like copper, aluminum, or carbon, whereas high resistance, such as 1M or more, indicates a poor conductor made of insulating materials like glass, porcelain, or plastic.

A Material whose resistance is somewhere between that of a good conductor and that of a good insulator, such as silicon or germanium are called Semiconductor. They are typically used to manufacture diodes and transistors etc.

Resistance might be linear or non-linear, but it can never be negative. Linear resistance follows Ohm's Law because the voltage across the resistor is proportional to the current flowing through it.

Non-linear resistance defies Ohm's Law by exhibiting a voltage drop across it that is proportional to some power of the current. Resistance is pure and is unaffected by frequency, with its AC impedance equal to its DC resistance and so cannot be negative. Remember that resistance is never negative and is always positive. A resistor is a passive circuit component which is unable to deliver power or store energy. Instead, resistors absorb power which can be occurred as heat.

It is sometimes much easier to use the reciprocal of resistance (1/R) rather than resistance (R) itself for very low quantities of resistance, such as milli-ohms (mΩ). Conductance (G) is the reciprocal of resistance and characterizes a conductor's or device's ability to conduct electricity. To put it another way, the ease with which current flows.

A good conductor, like copper, has a high conductance value, whereas a bad conductor, like wood, has a low conductance value. Siemen (S) is the standard unit of measurement for conductance.

Conductance's unit is mho (ohm spelt backward), which is denoted by an inverted Ohm sign ℧. Power can also be stated using conductance as: $p = i^2/G = v^2G$.

The correlation between voltage (V) and current (I) in a circuit with constant resistance (R) generates a straight line i-v correlation with slope equal to the resistance value, as shown.

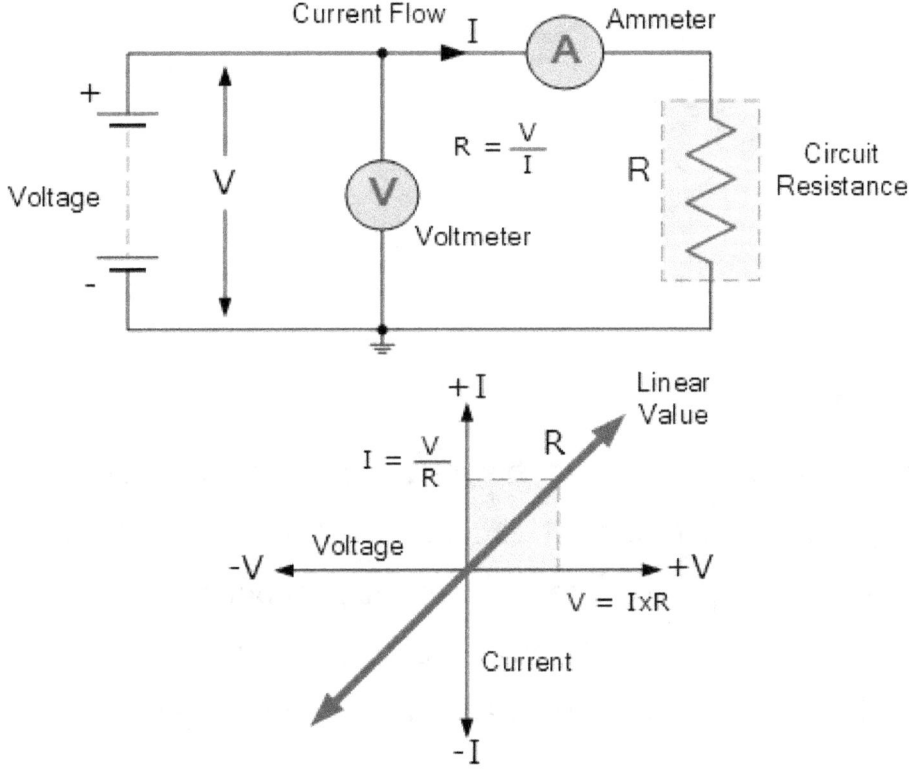

CHAPTER-2: OHM'S LAW AND POWER

Ohm's Law: The Relationship Between Voltage, Current and Resistance

Ohms Law is a mathematical equation which defines the relationship between Voltage, Current, and Resistance within electrical circuits. According to Ohm's Law: V = I*R.

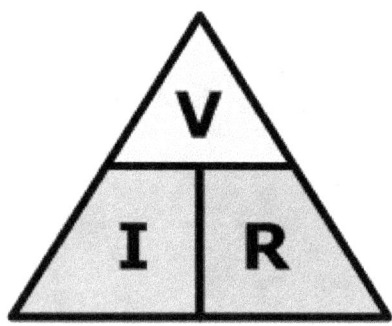

Ohms Law and Power

German physicist Georg Ohm first discovered the relationship between Voltage, Current and Resistance in any DC electrical circuit. According to Ohm's law, current flowing through a fixed linear resistance is directly proportional to the voltage applied across it, and also inversely proportional to the resistance in a constant temperature. This relationship between the Voltage, Current and Resistance is as follows"

$$Current, (I) = \frac{Voltage, (V)}{Resistance, (R)} \ in \ Amperes, (A)$$

We can calculate Voltage, (V) as follows,

V (volts) = I (amps) x R (Ω)

Current, (I) can be calculate as

I (amps) = V (volts) ÷ R (Ω)

Resistance, (R) can be found as follows,

R (Ω) = V (volts) ÷ I (amps)

Ohms Law in Triangle

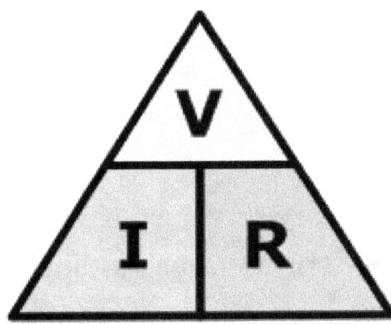

If we rearrange Ohm's Law, it will provide following arrangements of the same equation:

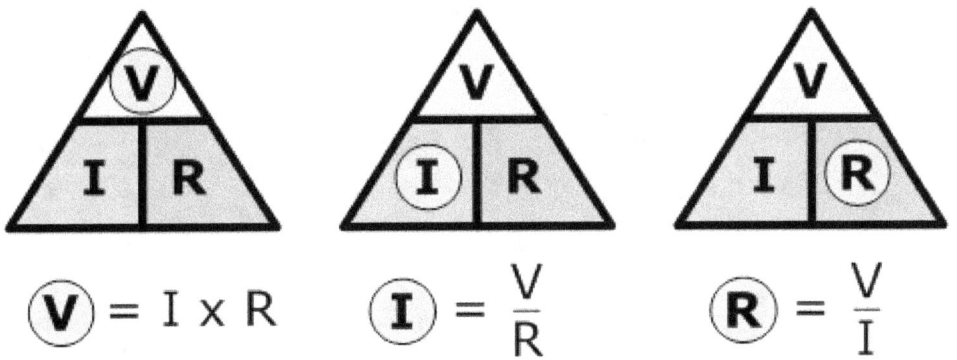

If 1V applied to 1Ω resistor, 1A current will flow which means current flows through any electrical circuit is directly proportional to the voltage across it (I α V).

Power

Power (P) is defined as the rate at which energy is transferred. It is converted to other form like motion, heat, or an electromagnetic field. The standard unit of Power is the watt (W). 1 kilowatt (kW) = 1000 W. 1 W is the power resulting from an energy dissipation, conversion, or storage process equivalent to one joule per second. According to Ohm's law, power can be calculated as follows:

P (watts) = V (volts) x I (amps)

Or, P (watts) = V^2 (volts) ÷ R (Ω)

Or, P (watts) = I^2 (amps) x R (Ω)

Now, if we superimpose these three quantities in a triangle shape, it will indicate power at the top and current and voltage at the bottom. It's known as Power Triangle. This arrangement denotes the actual position of each quantity within the Ohms law power formulas.

Power Triangle

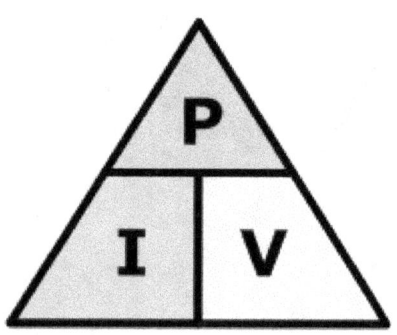

If we rearrange P, V and I as per Ohms Law, it will provide following combinations of the same equation for calculating individual values of P, V and I,

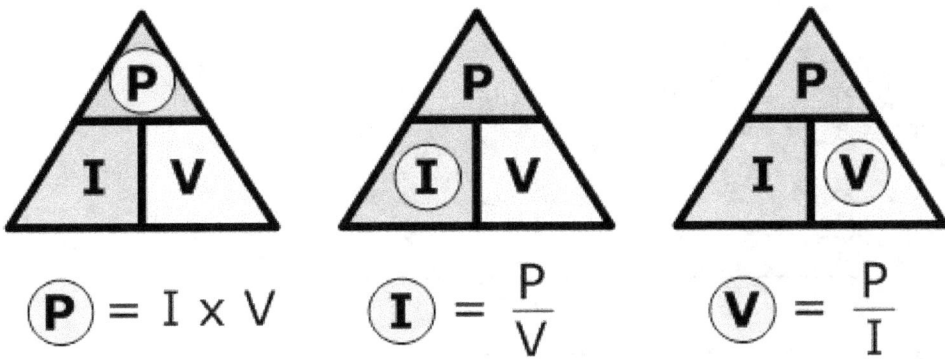

Ohm's Law in Pie Chart

Ohms Law pie can be rearranged as per following pie chart:

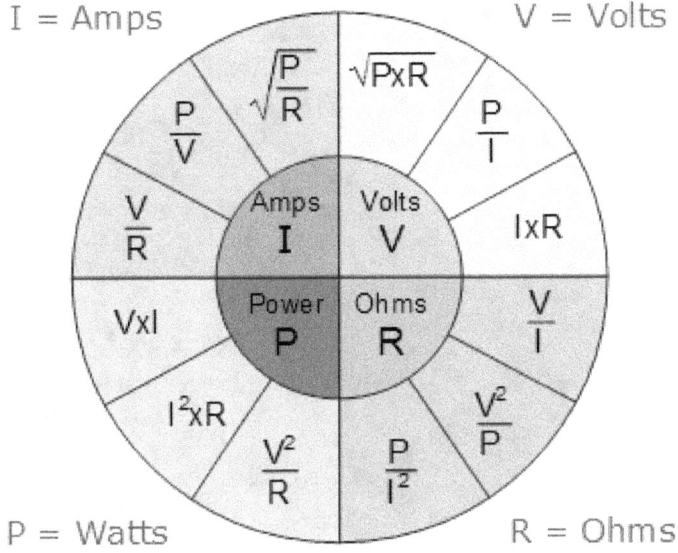

Ohm's Law in Matrix Table

Known Values	Resistance (R)	Current (I)	Voltage (V)	Power (P)
Current & Resistance	---	---	$V = I \times R$	$P = I^2 \times R$
Voltage & Current	$R = \dfrac{V}{I}$	---	---	$P = V \times I$
Power & Current	$R = \dfrac{P}{I^2}$	---	$V = \dfrac{P}{I}$	---
Voltage & Resistance	---	$I = \dfrac{V}{R}$	---	$P = \dfrac{V^2}{R}$
Power & Resistance	---	$I = \sqrt{\dfrac{P}{R}}$	$V = \sqrt{P \times R}$	---
Voltage & Power	$R = \dfrac{V^2}{P}$	$I = \dfrac{P}{V}$	---	---

Example-1 of Ohm's Law:

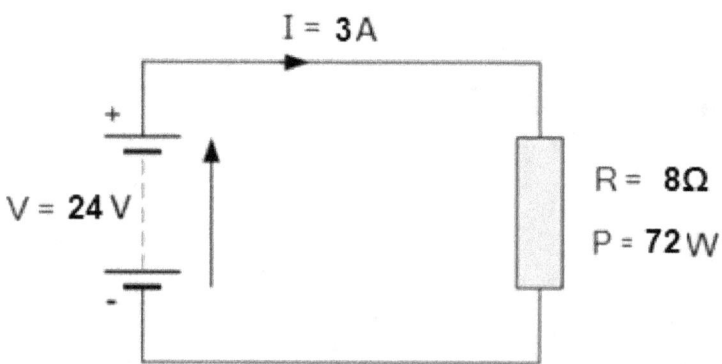

Voltage (V) = I x R = 3 x 8Ω = 24V

Current (I) = V ÷ R = 24 ÷ 8Ω = 3A

Resistance (R) = V ÷ I = 24 ÷ 3 = 8Ω

Power (P) = V x I = 24 x 3 = 72W

Power can only be found if both voltage and current exist in a circuit. For example, in an open-circuit, voltage exists but no current flow that is I = 0. As a result, no power will be dissipated in the circuit. Similarly, in a short-circuit, flow of current exist, but there is no voltage that is V = 0. Also, in this case, there will be no power dissipation.

Energy

Energy is the consumption of power in a time duration. Therefore, if we know how much power, in Watts is being consumed and the time, in seconds for which it is used, we can calculate the total energy used in watt-seconds. That is, Energy = Power x Time and Power = Voltage x Current. So, power is associated with energy. The unit of energy is the watt-seconds or *joules*.

Electrical Energy = Power (W) × Time (s)

CHAPTER-3: ELECTRICAL UNITS OF MEASURE

Power is the transfer of energy in a given rate. When one joule of work is either absorbed or delivered at a constant rate of one second, then the corresponding power will be equivalent to one watt. That is 1Joule/sec = 1Watt. Simply, power is the rate of doing work or the transferring of energy. Energy consumption is measured as Kilowatt-hours (kWh).

A "Unit of Electricity" is defined as the amount of electricity consumed by a device rated at 1000 watts in one hour. This is what the utility meter measures, and this is what we, as consumers, buy from our electricity suppliers when we get our bills. Kilowatt-hours are the standard units of energy used by our home's electricity meter to calculate how much electricity we use and thus how much we pay. So, if you turn on an electric fire with a 1000 Watt heating element and leave it on for an hour, you will have used 1 kWhr of electricity.

The total amount of electricity used would be exactly the same, or 1kWhr, if two electric fires with 1000 watt elements were turned on for 30 minutes each. Thus, using 1000 watts for one hour uses the same amount of energy as using 2000 watts (twice as much) for 30 minutes (half the time). Therefore, it would take 10 hours for a 100 Watt light bulb to consume 1 kWhr, or one unit, of electrical energy (10 x 100 = 1000 = 1kWhr).

CHAPTER-4: KIRCHHOFF'S CIRCUIT LAW

Kirchhoff's Circuit Laws define a set of basic network laws and theorems for the voltages and currents in a circuit, allowing us to solve complex circuit problems. We know that when two or more resistors are connected in series, parallel, or combinations of both, a single equivalent resistance (RT) can be establish, and that these circuits follow Ohm's Law.

But, in complex circuits like bridges or T networks, we can't always rely on Ohm's Law to determine the voltages and currents circulating within the circuit. Therefore, we require specific rules to get the circuit equations for these types of calculations, and we can use Kirchhoff's Circuit Law to do so.

Gustav Kirchhoff, a German physicist, established a set of rules or laws dealing with current and energy conservation in electrical circuits in 1845. Kirchhoff's Circuit Laws, with one of Kirchhoff's laws dealing with the current flowing around a closed circuit, Kirchhoff's Current Law, (KCL), and the other dealing with the voltage sources present in a closed circuit, Kirchhoff's Voltage Law, (KVL).

Kirchhoff's Current Law (KCL) states that "the total current or charge entering a junction or node is precisely equal to the charge leaving the node because it has nowhere else to go and no charge is lost within the node."

To put it another way, the algebraic sum of all currents entering and leaving a node must be zero, $I_{(exiting)} + I_{(entering)} = 0$. This concept of Kirchhoff is usually referred as the Conservation of Charge.

Kirchhoff's Current Law

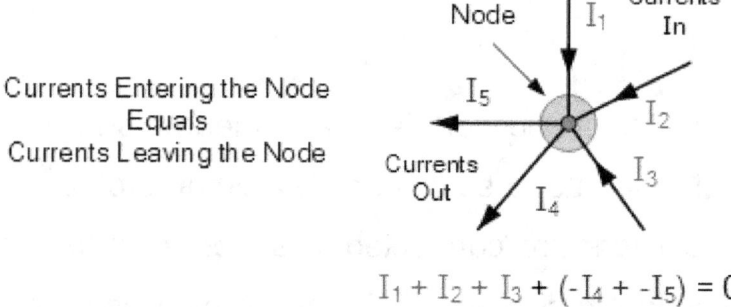

$$I_1 + I_2 + I_3 + (-I_4 + -I_5) = 0$$

The three currents which enter the node, I1, I2, and I3, are all positive in value, while the two currents that leave the node, I4 and I5, are both negative. Therefore, the equation can be expressed as I1 + I2 + I3 – I4 – I5 = 0.

A connection or junction of two or more current carrying paths or elements like cables and components, is referred to as a node in an electrical circuit. A closed circuit path must also exist for current to flow in and out of a node. When analyzing parallel circuits, we can use Kirchhoff's current law.

Kirchhoff's Voltage Law (KVL)

Kirchhoff's Voltage Law (KVL) states that "the total voltage around any closed loop network is equal to the sum of all the voltage drops within the same loop," and also equal to zero. To put it another way, the algebraic sum of all voltages in the loop must be zero. Kirchhoff's concept is familiar as the Conservation of Energy.

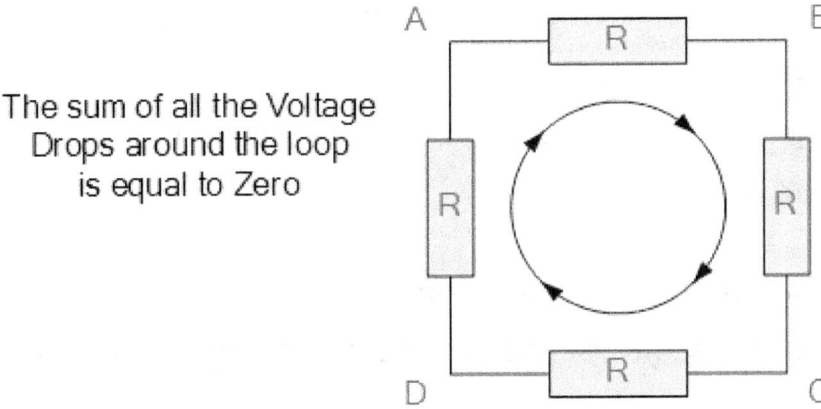

The sum of all the Voltage Drops around the loop is equal to Zero

$$V_{AB} + V_{BC} + V_{CD} + V_{DA} = 0$$

Beginning at any point in the loop, resume in the same direction, indicating the direction of all voltage drops (positive or negative), and returning to the same starting point. It's crucial to keep going in the same direction, whether clockwise or anti-clockwise, or the final voltage sum won't be zero. When studying series circuits, we can use Kirchhoff's voltage law.

When using Kirchhoff's Circuit Laws to analyze DC or AC circuits, a number of terms are used to define the parts of the circuit under investigation, including node, paths, branches, loops, and meshes. Since these terms are often used in circuit analysis, it's crucial to know them.

Common DC Circuit Theory Terms:

- **Circuit:** A closed loop path through which current flows.
- **Path:** A single line of connecting elements or sources.
- **Node:** A node is a junction, connection, or terminal within a circuit which connects or joins two or more circuit elements, providing a connection point between two or more branches. A dot represents a node.
- **Branch:** A branch is a single or group of components connected between two nodes like resistors or a source.

- **Loop:** A loop in a circuit is a simple closed path in which no circuit element or node appears more than once.
- **Mesh:** A mesh is a single closed loop series path that has no other paths in it. Within a mesh, there are no loops.

It should be noted here that when same current flows through components, it's called series connection and when same voltage applied across them, it's called parallel connection.

A Typical DC Circuit

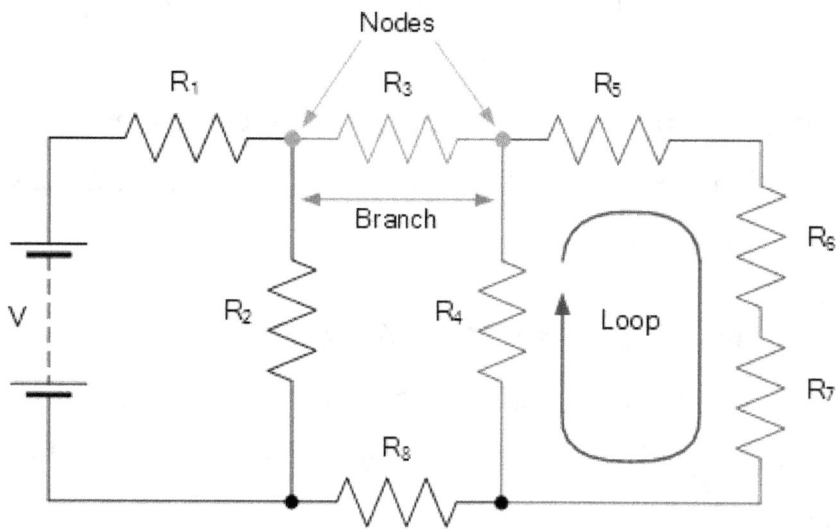

Example-1 of Kirchhoff's Circuit Law

Find the current flowing in the 20Ω Resistor, R_3

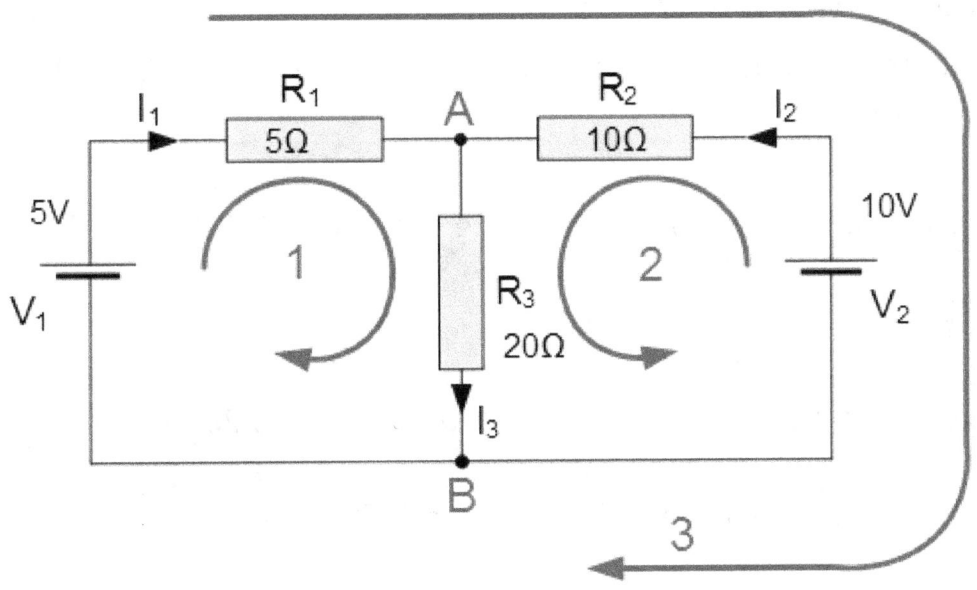

This circuit contains 3 branches, 2 nodes (A and B) and 2 independent loops.

Applying Kirchhoff's Current Law (KCL), the equations become as follows:

Node A: $I_1 + I_2 = I_3$

Node B: $I_3 = I_1 + I_2$

Applying Kirchhoff's Current Law (KCL), the equations become as follows:

Loop 1: $5 = R_1 I_1 + R_3 I_3 = 5I_1 + 20I_3$

Loop 2: $10 = R_2 I_2 + R_3 I_3 = 10I_2 + 20I_3$

Loop 3: $5 - 10 = 5I_1 - 10I_2$

Since I_3 is the sum of $I_1 + I_2$, we can rearrange the equations as follows:

Equation No. 1: $5 = 5I_1 + 20(I_1 + I_2) = 25I_1 + 20I_2$

Equation No. 2: $10 = 10I_2 + 20(I_1 + I_2) = 20I_1 + 30I_2$

Now, we have two "Simultaneous Equations" which can be subtracted to get the values of I1 and I2. Replacement of I_1 in terms of I_2 provides the value of I_1 as -0.143 Amps.

Replacement of I_2 in terms of I_1 provides the value of I_2 as +0.429 Amps

Since: $I_3 = I_1 + I_2$

Current flowing through resistor R_3 will be: -0.143 + 0.429 = 0.286 Amps

Voltage across the resistor R_3 will be: 0.286 x 20 = 5.72 volts

Application of Kirchhoff's Circuit Laws

These two laws allow the Currents and Voltages in a circuit to be found, ie, the circuit is called "Analyzed", and the basic method for using **Kirchhoff's Circuit Laws** is as follows:

- Let's, all voltages and resistances are specified. (If not tag them V1, V2,... R1, R2, etc.)
- Consign a current to each branch or mesh (clockwise or anticlockwise)
- Tag each branch with a branch current. (I1, I2, I3 etc.)
- Apply Kirchhoff's first law equations for each node.
- Apply Kirchhoff's second law equations for each of the independent loops of the circuit.
- Apply Linear simultaneous equations as required to get the unknown currents.

We can use loop analysis to calculate the currents in each independent loop in addition to using Kirchhoff's Circuit Law to calculate the various voltages and currents circulating around a linear circuit, which helps to reduce the amount of mathematics required by using only Kirchhoff's laws.

CHAPTER-5: MESH CURRENT ANALYSIS

Mesh Current Analysis is a method for determining the currents circulating around a loop or mesh in any circuit's closed path. While Kirchhoff's Laws provide the foundation for analyzing any complex electrical circuit, there are several ways to improve on this method, such as using Mesh Current Analysis or Nodal Voltage Analysis, which reduces the amount of math required, which can be a significant benefit when dealing with large networks. Consider the following electrical circuit as an example.

Circuit for Analyzing Mesh Current

Let's consider following circuit

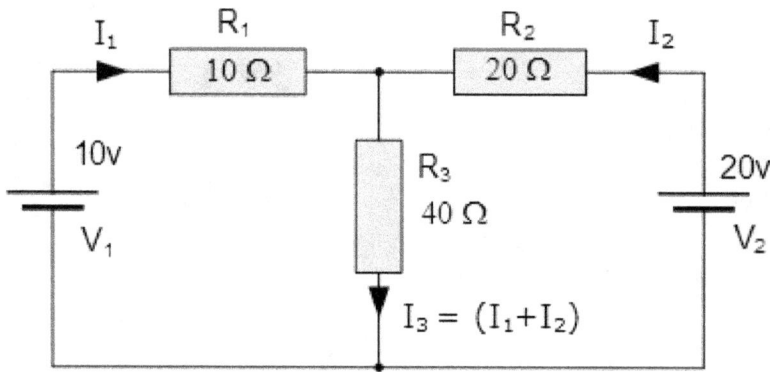

We can minimize the involvement of math for analyzing the circuit applying Kirchhoff's Current Law equations to find the currents, I_1 and I_2 flowing in the two resistors. Then there is no need to calculate the current I_3 as its just the sum of I_1 and I_2. Therefore, Kirchhoff's second voltage law just becomes:

Equation No 1: $10 = 50I_1 + 40I_2$

Equation No 2: $20 = 40I_1 + 60I_2$

In this way, one line of math's calculation has been saved.

Mesh Current Analysis

Above circuit can be solved easily by applying **Mesh Current Analysis** or **Loop Analysis** which is also often known as **Maxwell's Circulating Currents** method. Each "closed loop" require to be labeled with a circulating current instead of labelling currents of the branch.

According to thumb's rule, only label inside loops in a clockwise direction with circulating currents as the aim is to cover all the elements of the circuit at least once. Any required branch current may be determined from the appropriate loop or mesh currents as before applying Kirchhoff's method.

For example: $i_1 = I_1$, $i_2 = -I_2$ and $I_3 = I_1 - I_2$

Kirchhoff's voltage law equation can be written in the same way as before to solve them but the benefit of this method is that it confirms that the information obtained from the circuit equations is the minimum required to solve the circuit as the information is more general and can simply be put into a matrix form.

Let's consider following circuit:

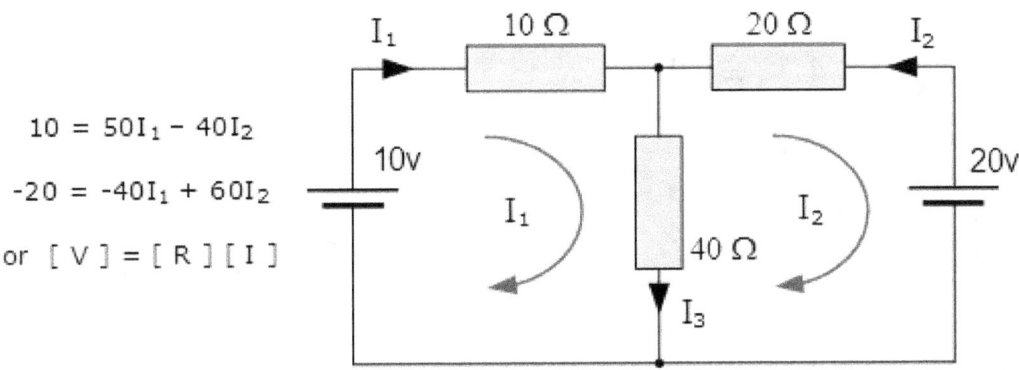

$10 = 50I_1 - 40I_2$

$-20 = -40I_1 + 60I_2$

or $[V] = [R][I]$

By applying a single mesh impedance matrix Z, these equations can be solved pretty quickly. Every component ON the principal diagonal will be "positive" and is the total impedance of each mesh. While, each element OFF the principal diagonal will either be "zero" or "negative" and denotes the circuit element connecting all the applicable meshes.

During dealing with matrices, we have to understand that, for the division of two matrices it is the same as multiplying one matrix by the inverse of the other as follows:

$$[V] = [I] \times [R] \quad or \quad [R] \times [I] = [V]$$

$$\begin{bmatrix} 50 & -40 \\ -40 & 60 \end{bmatrix} \times \begin{bmatrix} I_1 \\ I_2 \end{bmatrix} = \begin{bmatrix} 10 \\ -20 \end{bmatrix}$$

$$I = \frac{V}{R} = R^{-1} \times V$$

$$Inverse\ of\ R = \begin{bmatrix} 60 & 40 \\ 40 & 50 \end{bmatrix}$$

$$|R| = (60 \times 50) - (40 \times 40) = 1400$$

$$\therefore R^{-1} = \frac{1}{1400} \begin{bmatrix} 60 & 40 \\ 40 & 50 \end{bmatrix}$$

Getting the inverse of R, as V/R is the same as V x R[-1], we can now use it to determine two circulating currents.

$$[I] = [R^{-1}] \times [V]$$

$$\begin{bmatrix} I_1 \\ I_2 \end{bmatrix} = \frac{1}{1400} \begin{bmatrix} 60 & 40 \\ 40 & 50 \end{bmatrix} \times \begin{bmatrix} 10 \\ -20 \end{bmatrix}$$

$$I_1 = \frac{(60 \times 10) + (40 \times -20)}{1400} = \frac{-200}{1400} = -0.143 A$$

$$I_2 = \frac{(40 \times 10) + (50 \times -20)}{1400} = \frac{-600}{1400} = -0.429 A$$

Where:

[V]   gives the total battery voltage for loop 1 and then loop 2

[I]   states the names of the loop currents which we are trying to find

[R]   is the resistance matrix

[R^{-1}]   is the inverse of the [R] matrix

and this provides I_1 as -0.143 Amps and I_2 as -0.429 Amps

Since: $I_3 = I_1 - I_2$

So, the combined current of I_3 will be: -0.143 – (-0.429) = 0.286 Amps

This is the same value of 0.286 amps current.

Mesh Current Analysis in Brief

The best of all the circuit analysis methods is this "look-see" method of circuit which includes the basic process for solving **Mesh Current Analysis** equations. It is as follows:

- All the internal loops with circulating currents. (I_1, I_2, ...I_L etc) to be labelled.
- [L x 1] column matrix [V] to be written providing the sum of all voltage sources in each loop.
- [L x L] matrix, [R] for all the resistances in the circuit to be written as follows:
    - R_{11} = the total resistance in the first loop.
    - R_{nn} = the total resistance in the Nth loop.
    - R_{JK} = the resistance which directly joins loop J to Loop K.
- The matrix or vector equation [V] = [R] x [I] to be written where [I] is the list of currents to be found.

Apart from applying **Mesh Current Analysis**, node analysis also to be applied for calculating the voltages around the loops, again decreasing the amount of mathematics needed applying just Kirchoff's laws.

CHAPTER-6: NODAL VOLTAGE ANALYSIS

Unknown voltage drops around a circuit between different nodes can be determined by Nodal Voltage Analysis. It gives a common connection for two or more circuit components.

Nodal Voltage Analysis accompaniments the previous mesh analysis in that it is equally powerful and based on the same concepts of matrix analysis. As its name indicates, **Nodal Voltage Analysis** applies the "Nodal" equations of Kirchhoff's first law for determining the voltage potentials around the circuit.

Therefore, by adding together all these nodal voltages the net result will be equal to zero. If there are "n" nodes in the circuit, there will be "n-1" independent nodal equations and these alone are sufficient to define and hereafter solve the circuit.

Kirchhoff's first law equation to be written down in each node point which means that *"the currents entering a node are exactly equal in value to the currents leaving the node"* then express each current in terms of the voltage across the branch. For "n" nodes, one node will be utilized as the reference node and all the other voltages will be referenced or measured with respect to this common node. For example, consider following circuit:

Circuit for Nodal Voltage Analysis

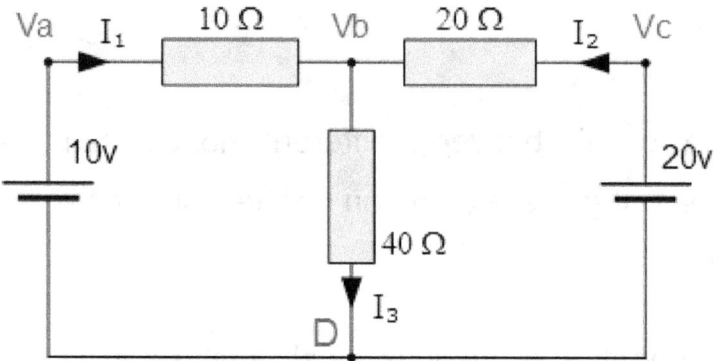

Here, node D is chosen as the reference node and the other three nodes are assumed to have voltages, Va, Vb and Vc with respect to node D. For example;

$$\frac{(V_a - V_b)}{10} + \frac{(V_c - V_b)}{20} = \frac{V_b}{40}$$

Since Va = 10v and Vc = 20v, Vb can be easily determined by:

$$\left(1 - \frac{Vb}{10}\right) + \left(1 - \frac{Vb}{20}\right) = \frac{Vb}{40}$$

$$2 = Vb\left(\frac{1}{40} + \frac{1}{20} + \frac{1}{10}\right)$$

$$Vb = \frac{80}{7} V$$

$$\therefore I_3 = \frac{2}{7} \text{ or } 0.286 Amps$$

again is the same value of 0.286 amps.

This is the easiest method of solving particular circuit. Usually, nodal voltage analysis is more suitable when there are a larger number of current sources around. The network is then described as: [I] = [Y] [V] where [I] are the driving current sources, [V] are the nodal voltages to be determined and [Y] is the admittance matrix of the network which functions on [V] for providing [I].

Nodal Voltage Analysis in Brief:

The basic method for solving **Nodal** Analysis equations is as follows:

- Inscribe the current vectors, let's consider currents into a node are positive. i.e., a (N x 1) matrices for "N" independent nodes.
- Inscribe the admittance matrix [Y] of the network where:
  Y_{11} = the total admittance of the first node.
  Y_{22} = the total admittance of the second node.
  R_{JK} = the total admittance joining node J to node K.
- In the network with "N" independent nodes, [Y] will be an (N x N) matrix and that Ynn will be positive and Yjk will be negative or zero value.
- The voltage vector will be (N x L) and will list the "N" voltages to be found.

We have observed that a number of theorems occur which basically the analysis of linear circuits.

CHAPTER-7: THEVENIN'S THEOREM

For changing a complex circuit into a modest equivalent circuit containing of a single resistance in series with a source voltage, an analytical method is applied which is known as Thevenin's theorem.

Thevenin's Theorem defines that any linear circuit having several voltages and resistances can be replaced by just one single voltage in series with a single resistance connected across the load. That means, any electrical circuit regardless of its complexity can easily be simplified to an equivalent two-terminal circuit with just a single constant voltage source in series with a resistance (or impedance) connected to a load as shown below. Thevenin's Theorem is particularly applicable in the circuit analysis of power or battery systems and other interconnected resistive circuits where it will have an effect on the adjoining part of the circuit.

Thevenin's equivalent circuit

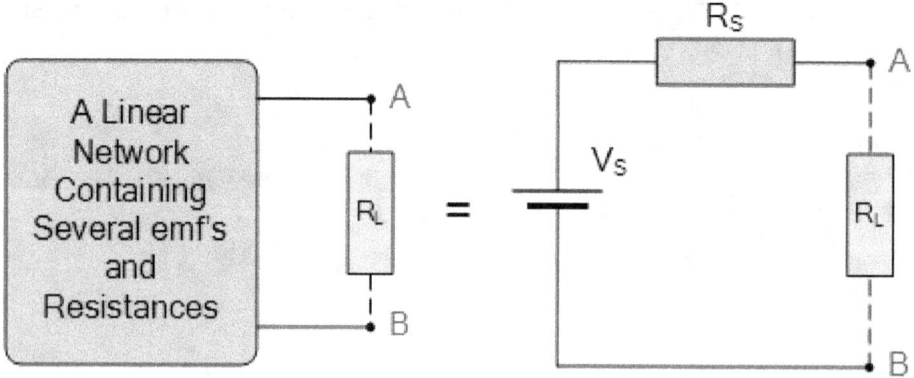

Any complex "one-port" network consisting of various resistive circuit elements and energy sources can be substituted by one single equivalent resistance R_s and one single equivalent voltage V_s in terms of the load resistor R_L. R_s is the source resistance looking back into the circuit, and V_s denotes the open circuit voltage at the terminals. Consider the following circuit as an example.

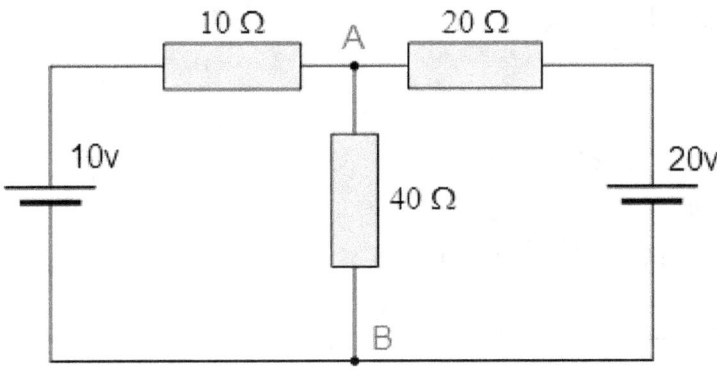

To begin analyzing the circuit, we must remove the center 40Ω load resistor connected across terminals A-B, as well as any internal resistance related with the voltage source (s). This is accomplished by shorting out all voltage sources connected to the circuit, resulting in v = 0, or by open circuiting all attached current sources, resulting in I = 0. The rationale for this is that we want to have an optimal voltage or current source for circuit examination.

Calculating the total resistance looking back from terminals A and B with all voltage sources shorted yields the value of the equivalent resistance, R_s. The following circuit results.

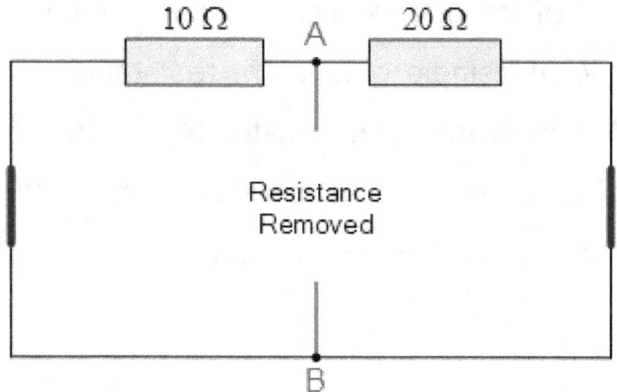

Determine the Equivalent Resistance (R_s)

10Ω Resistor in Parallel with the 20Ω Resistor

$$R_T = \frac{R_1 \times R_2}{R_1 + R_2} = \frac{20 \times 10}{20 + 10} = 6.67\Omega$$

The voltage V_s is defined as the total voltage across terminals A and B when they are connected in an open circuit. That is, without the load resistor R_L.

Determine the Equivalent Voltage (V_s)

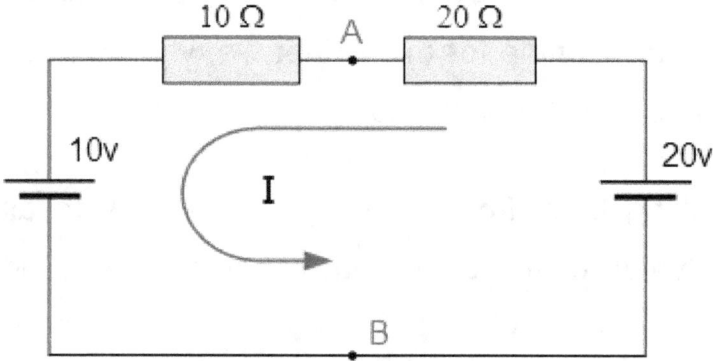

We must now reconnect the two voltages to the circuit, and because $V_S = V_{AB}$, the current traveling around the loop is computed as follows:

$$I = \frac{V}{R} = \frac{20v - 10v}{20\Omega + 10\Omega} = 0.33 \; amps$$

Since this current of 0.33 amps (330mA) is shared by both resistors, the voltage drop across the 20 resistor or the 10 resistor can be computed as follows:

V_{AB} = 20 − (20Ω x 0.33amps) = 13.33 volts.

or

V_{AB} = 10 + (10Ω x 0.33amps) = 13.33 volts, the same.

Thevenin's Equivalent circuit would then be composed of a series resistance of 6.67 and a voltage source of 13.33v. When we reconnect the 40 resistor to the circuit, we get:

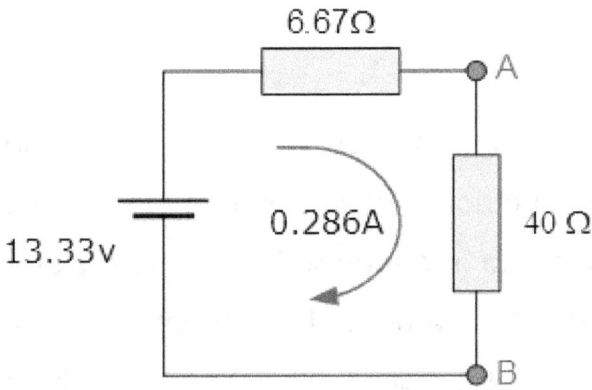

The current passing through the circuit is calculated as follows:

$$I = \frac{V}{R} = \frac{13.33v}{6.67\Omega + 40\Omega} = 0.286 \; amps$$

This is the same value of 0.286 amps that we discovered using Kirchhoff's circuit rule.

Thevenin's theorem is another sort of circuit analysis method that is very effective in the study of complex circuits that consist of one or more voltage or current sources and resistors arranged in the normal parallel and series connections.

Even though Thevenin's circuit theorem can be mathematically expressed in terms of current and voltage, it is less effective in larger networks than mesh current analysis or nodal voltage analysis because mesh or nodal analysis is typically required in any Thevenin exercise, so it might as well be used right away. Thevenin's equivalent circuits for transistors, voltage sources like batteries, etc., on the other hand, are extremely helpful when designing circuits.

Thevenin's Theorem in A Brief

As we've seen, Thevenin's theorem is another kind of circuit analysis technique that may be used to simplify any complex electrical network into a circuit that consists of a single voltage source (V_s) connected in series with a single resistor (R_s). This simple circuit acts electrically exactly the same as the intricate circuit it replaces when seen from terminals A and B. In other words, the i-v relationships at terminals A and B are the same. Thevenin's Theorem can be used to solve a circuit in the following manner:

1. Take out the load resistor RL or component which is concerned.
2. To determine RS, either short or open circuit all sources of voltage or current.
3. Use the standard circuit analysis techniques to find VS.
4. Determine the load resistor's current flow.

CHAPTER-8: NORTON'S THEOREM

Using Norton's theorem, a complex circuit can be reduced to a simple equivalent circuit that just has one resistance connected in parallel to a current source. As opposed to Norton, who parallels a steady current source with a single resistance in his circuit.

According to Norton's Theorem, any linear circuit with many energy sources and resistances can be replaced by a single constant current generator running in parallel with a single resistor.

As far as the load resistance, or R_L, is concerned, this single resistance, R_S, represents the value of the resistance looking back into the network with all current sources open circuited, and I_S, represents the short circuit current at the output terminals, as given below.

Norton's equivalent circuit

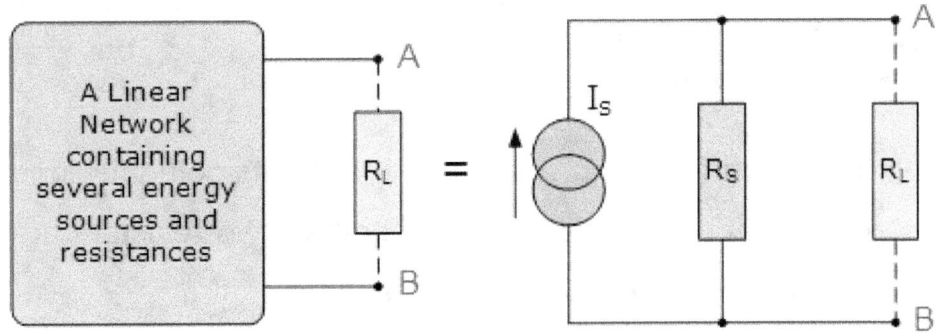

This "constant current" number represents the current that would flow if the two output terminals were shorted together while the source resistance was measured looking back into the terminals (the same as Thevenin's Theorem). Consider the preceding section's now-familiar circuit.

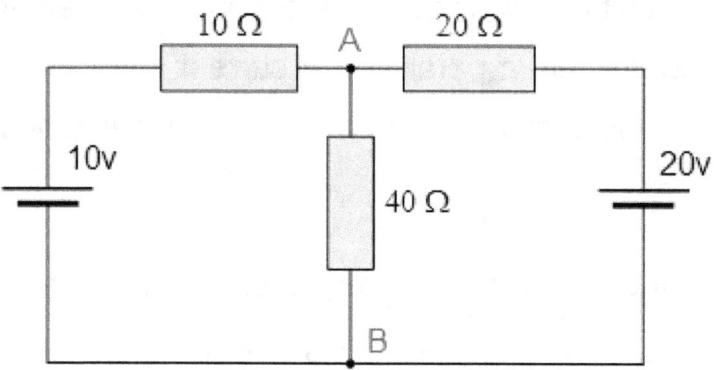

To determine the Norton's equivalent of the above circuit, disconnect the center 40Ω load resistor and short out terminals A and B, providing the following circuit.

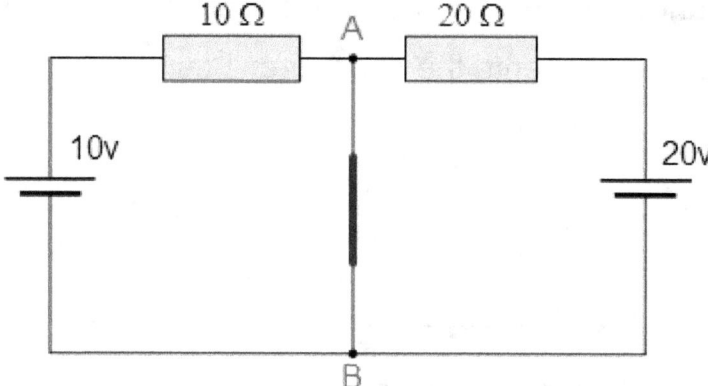

In order to determine the total short circuit current and the currents going through each resistor when the terminals A and B are shorted together, two resistors are connected in parallel across their two respective voltage sources.

with A-B Shorted Out

$$I_1 = \frac{10v}{10\Omega} = 1\,\text{amp}, \quad I_2 = \frac{20v}{20\Omega} = 1\,\text{amp}$$

$$\text{therefore,} \quad I_{\text{short-circuit}} = I_1 + I_2 = 2\,\text{amps}$$

The two resistors can be connected in parallel by shorting out the two voltage sources and opening up terminals A and B. By computing the combined resistance at terminals A and B, we can determine the value of the internal resistor R_s, which results in the circuit shown below.

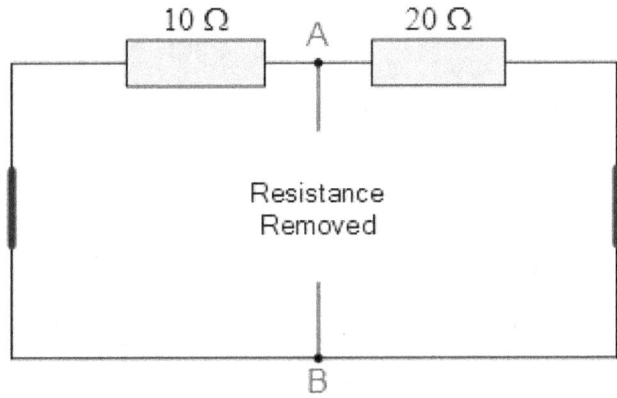

Determine the Equivalent Resistance (R_s)

10Ω Resistor in Parallel with the 20Ω Resistor

$$R_T = \frac{R_1 \times R_2}{R_1 + R_2} = \frac{20 \times 10}{20 + 10} = 6.67\Omega$$

After determining the short circuit current, I_s, and equivalent internal resistance, R_s, we may develop the Norton's equivalent circuit shown below.

Norton's equivalent circuit

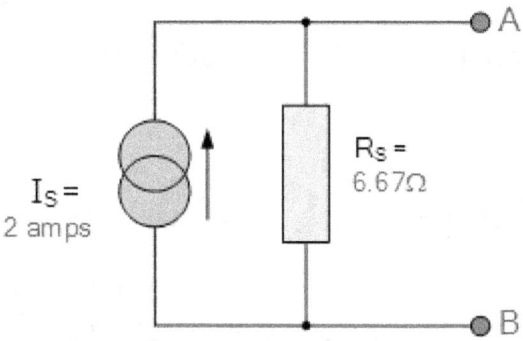

Now, let's solve with the original 40Ω load resistor connected across terminals A and B as follows:

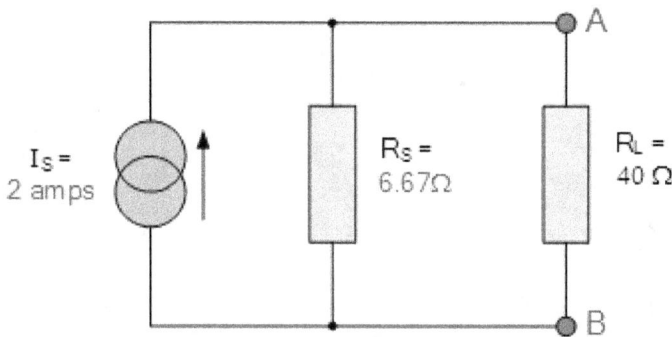

Here, two resistors are connected in parallel across the terminals A and B. Therefore, total resistance will be:

$$R_T = \frac{R_1 \times R_2}{R_1 + R_2} = \frac{6.67 \times 40}{6.67 + 40} = 5.72 \Omega$$

Now, voltage across the terminals A and B with the load resistor connected will be:

$$V_{A-B} = I \times R = 2 \times 5.72 = 11.44 \text{v}$$

Current flowing through 40Ω load resistor will be:

$$I = \frac{V}{R} = \frac{11.44}{40} = 0.286 \text{ amps}$$

Norton's Theorem in A Brief

We can solve a circuit by using Norton's Theorem following below mentioned steps:

1. Eliminate the load resistor R_L or component concerned.
2. Determine R_S by shorting all voltage sources or by open circuiting all the current sources.
3. Determine I_S by placing a shorting link on the output terminals A and B.
4. Determine the current flowing through the load resistor R_L.

When the load resistance and the source resistance are equal in a circuit, power supplied to the load is at its maximum.

CHAPTER-9: MAXIMUM POWER TRANSFER

When the resistive value of the load is equal to the internal resistance of the voltage sources, maximum power can be supplied. This is referred to as Maximum Power Transfer. In general, the source resistance, or impedance if inductors or capacitors are used, has a fixed value in Ohms.

When we connect a load resistance, R_L, across the power source's output terminals, the impedance of the load changes from open-circuit to short-circuit, causing the power absorbed by the load to become dependent on the impedance of the actual power source. The load resistance must then be "Matched" to the impedance of the power source in order to absorb the maximum amount of power. This is the foundation of Maximum Power Transfer.

Another useful circuit analysis method is the Maximum Power Transfer Theorem, which ensures that the maximum amount of power is dissipated in the load resistance when the value of the load resistance is exactly equal to the resistance of the power source. The power in the load is determined by the relationship between the load impedance and the internal impedance of the energy source. Consider the following circuit.

Thevenin's Equivalent Circuit

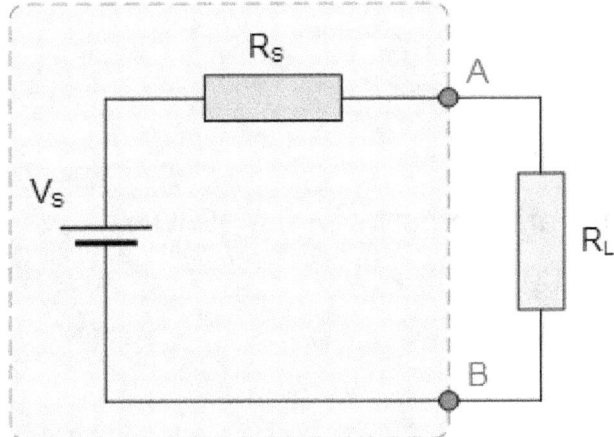

The maximum power transfer theorem states that "the maximum amount of power will be dissipated in the load resistance if it is equal in value to the Thevenin or Norton source resistance of the network supplying the power" in our Thevenin's equivalent circuit above.

In other words, if the load resistance resulting in the greatest power dissipation is equal to the equivalent Thevenin source resistance, then $R_L = R_S$; however, if the load resistance is less than or greater than the network's Thevenin source resistance, the dissipated power will be less than maximum.

Find the maximum power transfer in the following circuit by determining the value of the load resistance, R_L.

Example-1 of Maximum Power Transfer

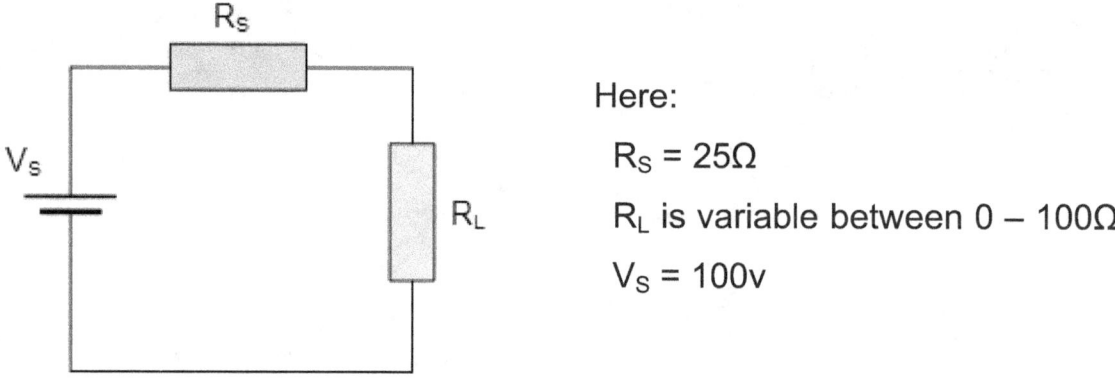

Here:

$R_S = 25\Omega$

R_L is variable between $0 - 100\Omega$

$V_S = 100v$

Applying Ohm's Law equations, it becomes:

$$I = \frac{V_S}{R_S + R_L} \quad \text{and} \quad P = I^2 R_L$$

Using below table, we can get the current and power in the circuit for different values of load resistance.

Table: Current against Power

R_L (Ω)	I (amps)	P (watts)	R_L (Ω)	I (amps)	P (watts)
0	4.0	0	25	2.0	**100**
5	3.3	55	30	1.8	97
10	2.8	78	40	1.5	94
15	2.5	93	60	1.2	83
20	2.2	97	100	0.8	64

We can plot a graph of load resistance, R_L against power, P for various values of load resistance using the information from the aforementioned table.

It should be noted here that power is zero for both a short-circuit and an open circuit (zero current condition) (zero voltage condition).

Power against Load Resistance Graph

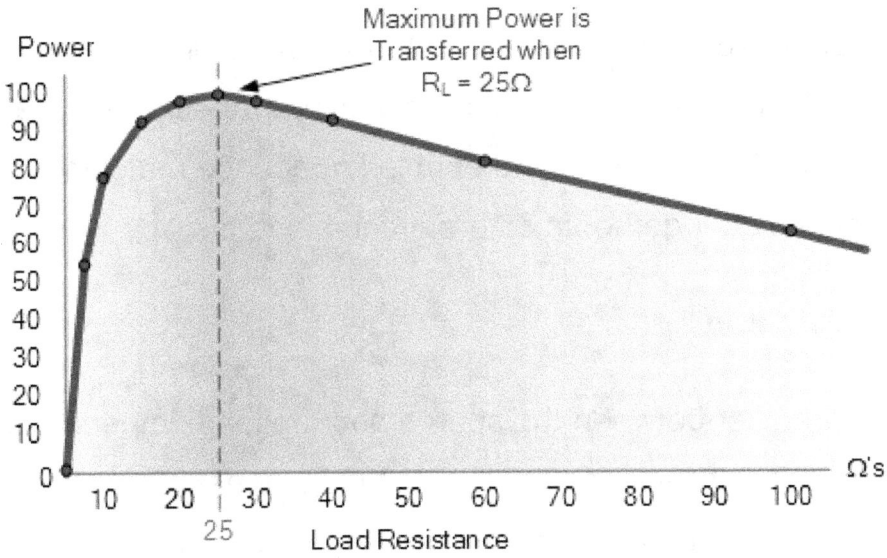

The maximum power transfer occurs in the load when the load resistance, RL, is equal in value to the source resistance, R_S, that is: $R_S = R_L = 25Ω$, as shown in the above table and graph. As a general rule, maximum power is transferred from an active device, such as a power source or battery, to an external device when the impedance of the external device precisely matches the impedance of the source. This is known as a "matched condition."

Impedance matching between an audio amplifier and a loudspeaker is a good illustration. The nominal input impedance, Z_{IN}, of the loudspeaker may only be specified as 8Ω, while the output impedance, Z_{OUT}, of the amplifier may be specified as being between 4Ω and 8Ω.

The amplifier will recognize the speaker as an 8Ω load if it is connected to the amplifier's output at that point. The amplifier can drive two 8Ω speakers in parallel in the same way that it can drive a single 4Ω speaker, and both connections fall within the amplifier's permitted output range. Inadequate impedance matching can cause a significant increase in power loss and heat generation.

However, given that an amplifier and loudspeaker have very different impedances, how could your impedance match them? For example, in PA (public address) systems, there are loudspeaker impedance matching transformers that can change impedances from 4Ω to 8Ω or 16Ω to enable impedance matching of numerous loudspeakers connected together in different combinations.

Transformer Impedance Matching

In order to maximize power transfer between the source and the load, impedance matching is a very helpful technique used in the output stages of amplifier circuits. To get the most sound output from an amplifier, signal transformers are used to match the loudspeakers' higher or lower output impedance value. These "matching transformers" for audio signals couple the load to the output of the amplifier as shown below.

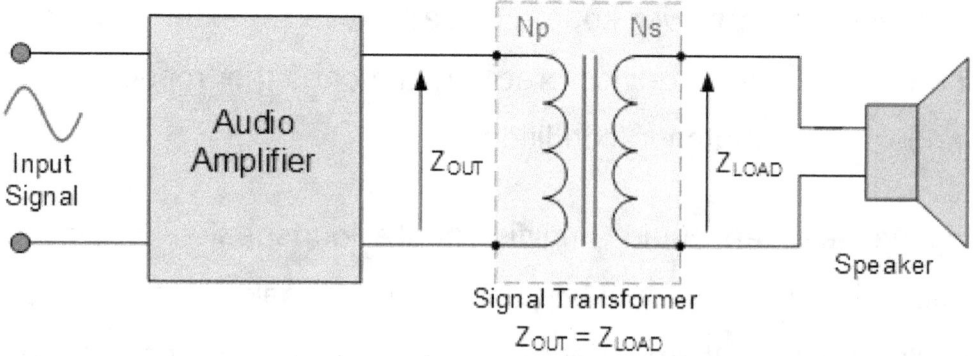

Even if the output impedance is different from the load impedance, the maximum power transfer can still be achieved. As a resistance on one side of the transformer changes to a different value on the other, this can be accomplished by using a suitable "turns ratio" on the transformer with the corresponding ratio of the load impedance, Z_{LOAD} to output impedance, Z_{OUT}. The following equation can be used to determine the maximum power transfer if the source impedance, Z_{OUT}, and the load impedance, Z_{LOAD}, are both entirely resistive.

$$Z_{out} = \left(\frac{N_P}{N_S}\right)^2 Z_{load}$$

Here, on the transformer side, N_P represents the number of primary turns and N_S represents the number of secondary turns. The output impedance can then be "matched" to the source impedance to achieve maximum power transfer by adjusting the value of the transformer's turns ratio. Have a look at following example:

Example-1: Calculate the turns ratio of the matching transformer required to provide maximum power transfer of the audio signal if an 8Ω loudspeaker is connected to an amplifier with an output impedance of 1000Ω. Assume the source impedance of the amplifier is Z_1, the load impedance is Z_2, and the turns ratio is N.

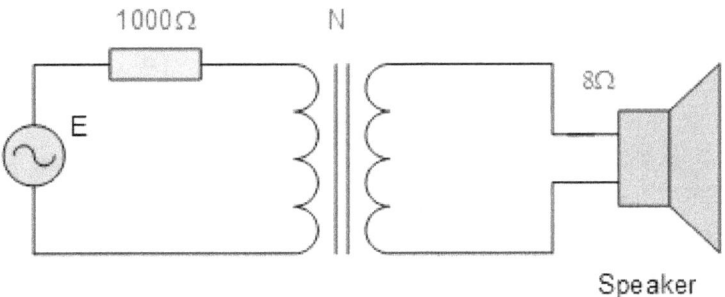

Speaker

$$Z_1 = N^2 Z_2 \therefore N = \sqrt{\frac{Z_1}{Z_2}}$$

therefore,

$$N = \sqrt{\frac{Z_1}{Z_2}} = \sqrt{\frac{1000}{8}} = 11.2:1$$

In low power amplifier circuits, small high frequency audio transformers are almost always regarded as ideal for simplicity, so any losses can be disregarded.

CHAPTER-10: STAR DELTA TRANSFORMATION

We can change the type of connection between impedances connected in a 3-phase configuration using the Star-Delta and Delta-Star transformations. Kirchhoff's Circuit Laws, mesh current analysis, and nodal voltage analysis techniques can now be used to solve simple series, parallel, or bridge type resistive networks, but in a balanced 3-phase circuit, we can use other mathematical techniques to simplify the analysis of the circuit and thereby reduce the amount of math required, which in and of itself is a good thing.

Standard three-phase circuits or networks are classified into two types based on how the resistances are connected: A Star connected network with the symbol of the letter Y (wye) and a Delta connected network with the symbol of a triangle Δ (delta).

If a three-phase, three-wire supply or even a three-phase load is connected in one configuration, it can be easily transformed or changed into an equivalent configuration of the other type using either the Star Delta Transformation or Delta Star Transformation process.

A resistive network with three impedances can be configured in a T or "Tee" configuration by connecting the impedances together, but the network can also be redrawn to create a star- or Y-shaped network, as is demonstrated below.

T-connected and Equivalent Star Network

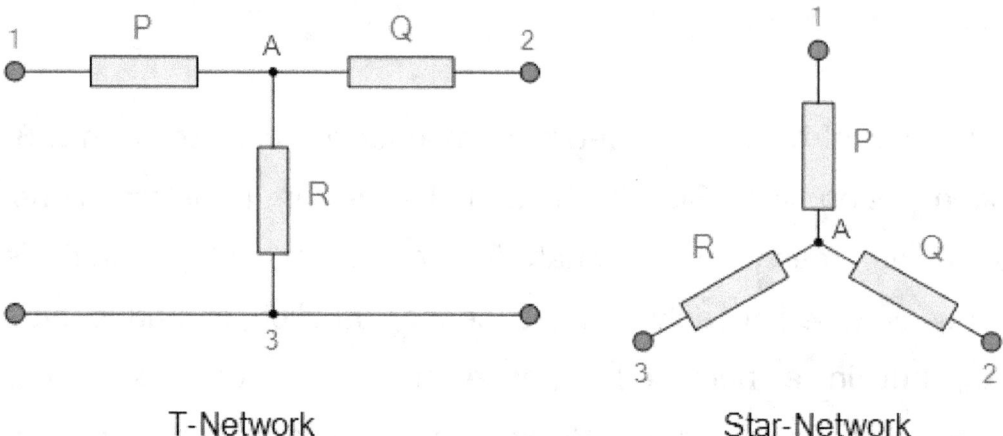

T-Network   Star-Network

The T resistor network above can be redrawn to create an electrically equivalent Star or Y type network, as we have already seen. However, as illustrated below, we can also change a Pi or type resistor network into an electrically equivalent Delta or Δ type network.

Pi-connected and Equivalent Delta Network

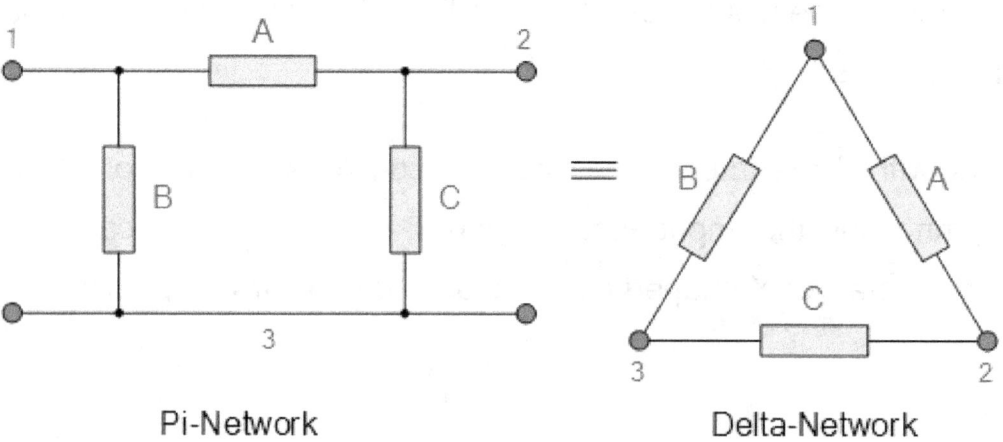

Pi-Network   Delta-Network

After defining what a Star and Delta connected network is, the transformation process can be used to convert the Y into an equivalent Δ circuit as well as to convert a Δ into an equivalent Y circuit. This procedure allows us to generate a mathematical relationship between the various resistors, yielding both a Star Delta Transformation and a Delta Star Transformation.

The three connected resistances (or impedances) can be changed by their equivalents measured between terminals 1-2, 1-3, or 2-3 for either a star-connected circuit or a delta-connected circuit using these circuit transformations.

However, since the internal voltages and currents of the star or delta networks differ, the resulting networks are only equivalent for voltages and currents that are external to the networks. Nevertheless, each network will use the same amount of power and have the same power factor with respect to the other networks.

Delta Star Transformation

A transformation formula is required to be derived for equating the various resistors to each other between the various terminals for converting a delta network to an equivalent star network. Let's see the following circuit:

Delta to Star Network

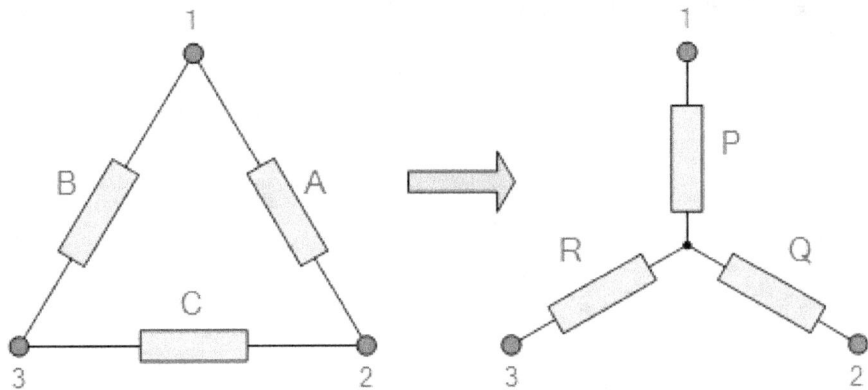

Let's compare the resistances between terminals 1 and 2.

$$P + Q = A \text{ in parallel with } (B + C)$$

$$P + Q = \frac{A(B + C)}{A + B + C} \quad \ldots EQ1$$

Resistance between the terminals 2 and 3.

$$Q + R = C \text{ in parallel with } (A + B)$$

$$Q + R = \frac{C(A + B)}{A + B + C} \quad \ldots EQ2$$

Resistance between the terminals 1 and 3.

$$P + R = B \text{ in parallel with } (A + C)$$

$$P + R = \frac{B(A + C)}{A + B + C} \quad \ldots EQ3$$

This now provides us three equations and taking equation 3 from equation 2 gives:

EQ3 - EQ2 = (P + R) - (Q + R)

$$P + R = \frac{B(A+C)}{A+B+C} - Q + R = \frac{C(A+B)}{A+B+C}$$

$$\therefore P - Q = \frac{BA+CB}{A+B+C} - \frac{CA+CB}{A+B+C}$$

$$\therefore P - Q = \frac{BA - CA}{A+B+C}$$

Then, re-writing Equation 1 will give us:

$$P + Q = \frac{AB + AC}{A+B+C}$$

After adding together equation 1 and the result above of equation 3 minus equation 2 provides:

(P - Q) + (P + Q)

$$= \frac{BA - CA}{A+B+C} + \frac{AB + AC}{A+B+C}$$

$$= 2P = \frac{2AB}{A+B+C}$$

From which provides us the final equation for resistor P as:

$$P = \frac{AB}{A+B+C}$$

After reviewing above math, it has been observed that resistor P in a Star network can be found as Equation 1 plus (Equation 3 minus Equation 2) or Eq1 + (Eq3 − Eq2).

Likewise, for determining resistor Q in a star network, is equation 2 plus the result of equation 1 minus equation 3 or Eq2 + (Eq1 − Eq3) and this provides us the transformation of Q as:

$$Q = \frac{AC}{A+B+C}$$

For determining resistor R in a Star network, is equation 3 plus the result of equation 2 minus equation 1 or Eq3 + (Eq2 − Eq1) and this gives us the transformation of R as:

$$R = \frac{BC}{A+B+C}$$

During transforming a delta network into a star network the denominators of all of the transformation formulas are the same: A + B + C, and which is the sum of all the delta resistances. For transforming any delta connected network to an equivalent star network, above transformation equations can be summarized as follows:

Delta to Star Transformations Equations

$$P = \frac{AB}{A+B+C} \quad Q = \frac{AC}{A+B+C} \quad R = \frac{BC}{A+B+C}$$

When the value of three resistors in the delta network are equal, then the resultant resistors in the equivalent star network will be equal to one third the value of the delta resistors. This provides each resistive branch in the star network a value of: R_{STAR} = 1/3*R_{DELTA} which is equivalent to: $(R_{DELTA})/3$

Delta – Star Example-1:

Transform the following Delta Resistive Network into an equivalent Star Network.

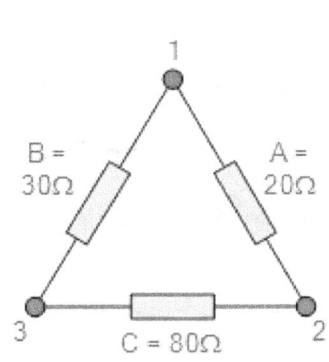

$$Q = \frac{AC}{A+B+C} = \frac{20 \times 80}{130} = 12.31\Omega$$

$$P = \frac{AB}{A+B+C} = \frac{20 \times 30}{130} = 4.61\Omega$$

$$R = \frac{BC}{A+B+C} = \frac{30 \times 80}{130} = 18.46\Omega$$

Star Delta Transformation

Simply said, the above transition in a star delta is reversed. We have seen that the resistor attached to one terminal is the product of the two delta resistances connected to that terminal when converting from a delta network to an equivalent star network. For instance, resistor P is the product of resistors A and B connected to terminal 1.

We may create a star delta transformation as indicated below by slightly altering the preceding formulas to find the transformation formulas for transforming a resistive star network into an analogous delta network.

Star to Delta Transformation

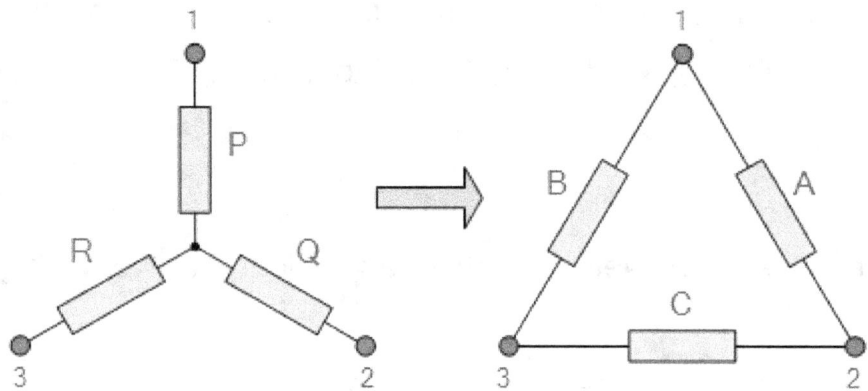

The total of all two-product combinations of resistors in the star network divided by the star resistor "immediately opposite" the delta resistor being searched determines the value of the resistor on any given side of the delta, network.

As an example, resistor A is described as follows:

$$A = \frac{PQ + QR + RP}{R}$$

the following with regard to terminal 3 and resistor B:

$$B = \frac{PQ + QR + RP}{Q}$$

with regard to terminal 2, using resistor C as described:

$$C = \frac{PQ + QR + RP}{P}$$

with respect to terminal 1.

Three independent transformation formulas are obtained by dividing each equation by the denominator value, and these formulas may be used to transform any Delta resistive network into an analogous star network, as given below:

$$A = \frac{PQ}{R} + Q + P \quad B = \frac{RP}{Q} + P + R \quad C = \frac{QR}{P} + Q + R$$

One more thing to consider when converting a star resistive network to an equivalent delta network. If all of the resistors in the star network have the same value, the resultant resistors in the equivalent delta network will be three times the value of the star resistors and equal, resulting in: $R_{DELTA} = 3*R_{STAR}$

Example-2 of Star – Delta:

Transform the following Star Resistive Network into an equivalent Delta Network.

$$A = \frac{QP}{R} + Q + P = \frac{180 \times 150}{60} + 180 + 150 = 780\Omega$$

$$B = \frac{RP}{Q} + R + P = \frac{60 \times 150}{180} + 60 + 150 = 260\Omega$$

$$C = \frac{QR}{P} + Q + R = \frac{180 \times 60}{150} + 180 + 60 = 312\Omega$$

We can transform one type of circuit connection into another type using both Star Delta Transformation and Delta Star Transformation in order to quickly analyze the circuit. These transformation methods work well for both star and delta circuits that contain impedances or resistances.

CHAPTER-11: VOLTAGE SOURCES

A voltage source is a component that produces a precise output voltage that, theoretically, remains constant regardless of the load current. The passive and active types of elements that make up an electrical or electronic circuit have been demonstrated. An active element is one that does not just consume energy but is able to continuously supply it to a circuit, such as a battery, generator, operational amplifier, etc.

The most important types of active circuit elements for us are those that supply electrical energy to the circuits or networks to which they are connected. These are referred to as "electrical sources," and there are two types of electrical sources: voltage sources and current sources. The current source is less common in circuits than the voltage source, but both are used and can be thought of as complementary.

A device that provides electrical power to a circuit in the form of a voltage source or a current source is known as an electrical supply, or simply "a source." Both kinds of electrical sources can be categorized as either direct (DC) or alternating (AC) sources, where a DC voltage is a constant voltage and an AC voltage is one that varies sinusoidally over time. So, for instance, batteries are DC sources, whereas your home's 230V mains outlet or wall socket is an AC source.

As mentioned earlier, electrical sources are a source of energy, but an intriguing feature of an electrical source is its capacity to transform non-electrical energy into electrical energy and vice versa.

A battery, for example, converts chemical energy into electrical energy, whereas a DC generator or an AC alternator converts mechanical energy into electrical energy. Renewable technologies can convert solar, wind, and wave energy into electrical or thermal energy. However, in addition to converting energy from one source to another, electrical sources can also deliver or absorb energy, allowing it to flow in both directions.

The I-V characteristics of an electrical source are another important feature that defines its operation. The I-V characteristic of an electrical source can provide us with a very nice pictorial description of the source, as shown as a voltage source and a current source.

Electrical Sources

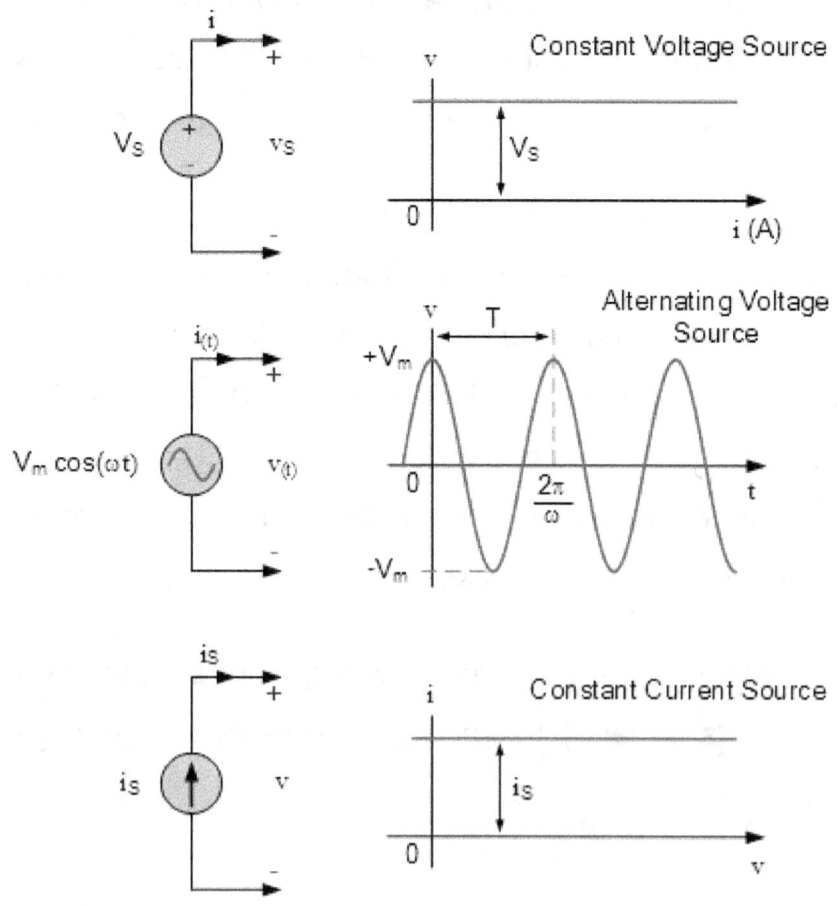

Electrical sources can be classified as independent (ideal) or dependent (controlled), that is, whose value depends on another voltage or current within the circuit, which can be either constant or time-varying.

Electrical sources are frequently considered to be "ideal" when discussing circuit laws and analysis; that is, the source is ideal because it could theoretically deliver an infinite amount of energy without loss, having characteristics that are represented by a straight line. However, there is always a resistance in actual or practical sources that is either connected in parallel for a current source or series for a voltage source, affecting the source's output.

Voltage Source

An electrical circuit can flow current by creating a potential difference (voltage) between two points by using a voltage source, such as a battery or generator. Keep in mind that voltage can exist independently of current. The most typical voltage source for a circuit is a battery, and the voltage that appears across the source's positive and negative terminals is referred to as the terminal voltage.

Ideal Voltage Source

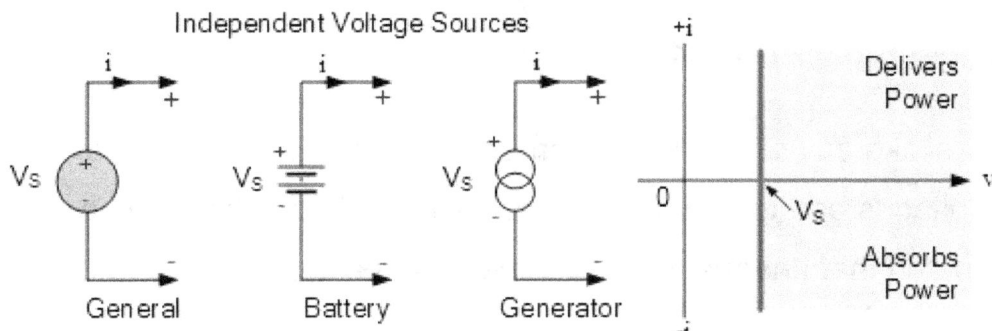

A two terminal active element that can supply and maintain the same voltage (v) across its terminals regardless of the current I flowing through it is referred to as an ideal voltage source.

In other words, the ideal voltage source will always produce a constant voltage, producing an I-V characteristic that looks like a straight line, regardless of the amount of current supplied.

An ideal voltage source is then referred to as an Independent Voltage Source because its voltage is determined solely by the value of the source and not by the amount or direction of the current flowing through it.

An automobile battery, for instance, has a 12V terminal voltage that, as long as the current flowing through it does not increase, remains constant, supplying power in one direction to the vehicle and absorbing power in the other as the battery charges.

A dependent voltage source, also known as a controlled voltage source, delivers a voltage supply whose magnitude is influenced by the voltage across or current passing through another circuit element. Many electronic devices, including transistors and operational amplifiers, use dependent voltage sources, which are denoted by a diamond shape and serve as equivalent electrical sources.

Connecting Voltage Sources Together

Ideal voltage sources, like any other circuit element, can be connected in parallel or series. Parallel voltages add together, whereas series voltages have the same value. It should be noted that unequal ideal voltage sources cannot be connected in parallel.

Voltage Sources in Parallel

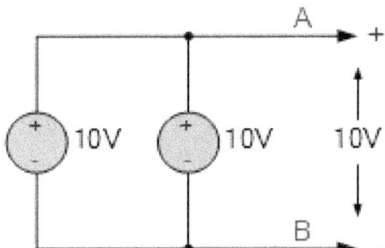

Ideal voltage sources can be connected in parallel as long as they have the same voltage value, even though this is not the best practice for circuit analysis. Two 10 volt voltage sources are combined in this example to produce 10 volts between terminals A and B.

In a perfect world, terminals A and B would be connected to a single 10 volt voltage source. Connecting ideal voltage sources that, as shown, have different voltage values or are short-circuited by an external closed loop or branch is not permitted and is not considered best practice.

Poorly Connected Voltage Sources

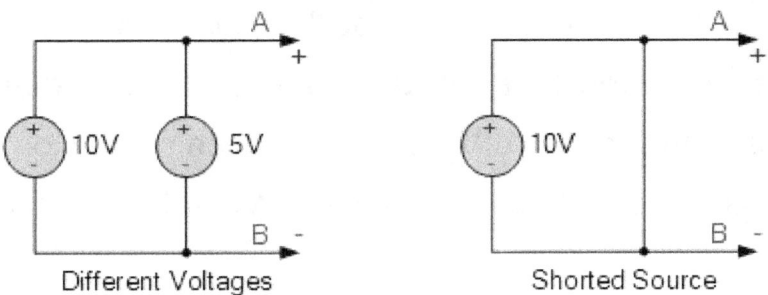

As long as there are other circuit components in between them to comply with Kirchoff's Voltage Law, KVL, it is possible to use voltage sources of different values when dealing with circuit analysis. Unlike voltage sources that are connected in parallel, ideal voltage sources of various values can be linked in series to create a single voltage source whose output is the algebraic addition or subtraction of the input voltages. Both series-aiding and series-opposing voltages, as shown, are possible for their connection.

Voltage Source in Series

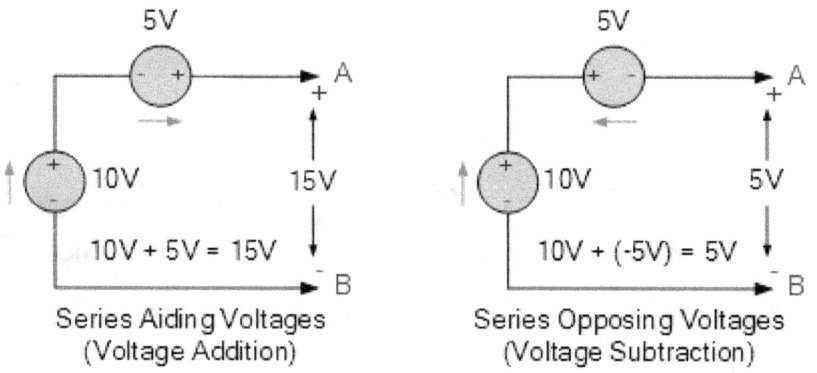

Series-connected voltage sources have their polarities connected so that the plus terminal of one is connected to the negative terminal of the next, allowing current to flow in the same direction. In the preceding example, the first circuit's two voltages of 10V and 5V can be added for a VS of 10 + 5 = 15V. As a result, the voltage between terminals A and B is 15 volts.

As shown in the second circuit above, series opposing voltage sources are series connected sources with their polarities connected so that the plus or negative terminals are connected together. As a result, the voltages are subtracted from one another. The second circuit's two voltages, 10V and 5V, are then subtracted, with the smaller voltage coming out ahead of the larger voltage. VS = 10 - 5 = 5V as a result.

The larger polarity of the voltage sources determines the polarity across terminals A and B, which in this example results in +5 volts when terminal A is positive and terminal B is negative. The net voltage across A and B will be zero if the series-opposing voltages are equal because one voltage cancels out the other. Any currents (I) will also be zero because current cannot flow in the absence of a voltage source.

Example No1 of Voltage Source

To supply a 100 Ohm load resistance, two series-connected ideal voltage sources at 6 and 9 volts each are connected. Calculate the source voltage (V_S), load current (I_R) through the resistor, and total power (P) that the resistor dissipates. Sketch the circuit.

$$V_S = V_1 + V_2 = 6 + 9 = 15V$$

$$I_R = \frac{V_S}{R} = \frac{15V}{100\Omega} = 150mA$$

$$P_R = I^2R = 0.15^2 \times 100 = 2.25W$$

In this way, V_S = 15V, I_R = 150mA or 0.15A, and P_R = 2.25W.

Practical Voltage Source

It has been observed that a perfect voltage source can supply a voltage that is unaffected by the current that passes through it, meaning that it always maintains the same voltage value. This concept might be effective for circuit analysis techniques, but in practice voltage sources behave a little differently because in a real-world voltage source, the terminal voltage actually decreases as the load current increases.

Since an ideal voltage source's terminal voltage does not change as the load current increases, this suggests that the internal resistance of an ideal voltage source is zero, or R_S = 0. In other words, it is a voltage source without resistors.

As they supply higher load currents, all voltage sources actually have a very low internal resistance that lowers their terminal voltage.

Since these two series-connected elements carry the same current as shown, internal resistance (R_S) from non-ideal or practical voltage sources, like batteries, has the same result as a resistance connected in series with an ideal voltage source.

Practical and Ideal Voltage Source

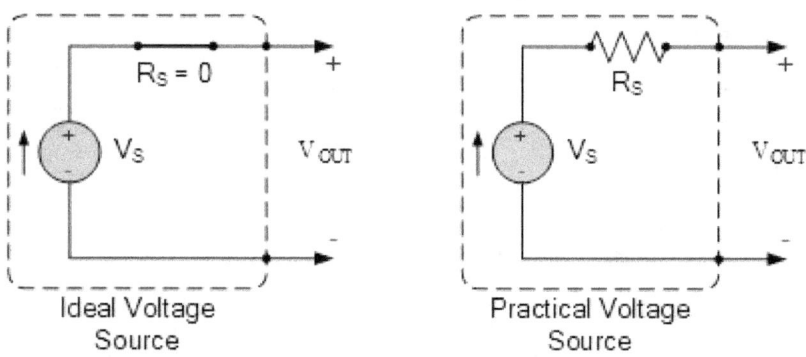

Ideal Voltage Source   Practical Voltage Source

Since Thevenin's theorem states that "any linear network containing resistances and sources of emf and current may be replaced by a single voltage source, V_S in series with a single resistance, R_S," you may have noticed that a practical voltage source closely resembles that of a Thevenin's equivalent circuit.

Keep in mind that the voltage source is suitable if the series source resistance is low. The voltage source is open-circuited when the source resistance is infinite. As the terminal voltage decreases with a rise in load current, the internal resistance, R_S, of every actual or practical voltage source, regardless of how modest, affects the source's I-V characteristic. This is due to the fact that R_S receives the same load current.

According to Ohm's law, a voltage drop occurs across a resistance when a current I runs through it. This voltage drop's value is expressed as i* R_S. The ideal voltage source, V_S, less the voltage drop across the resistor, $i*R_S$, will therefore equal V_{OUT}.

Keep in mind that R_S is equal to zero in the case of an ideal source voltage since there is no internal resistance; as a result, the terminal voltage is the same as V_S. Kirchoff's voltage law, KVL, thus states that the voltage sum around the loop is as follows: $V_{OUT} = V_S - i*R_S$.

The I-V characteristics of the actual output voltage can be obtained from this equation by plotting it. When the current I = 0, as shown, it will result in a straight line with a slope of $-R_S$ that intersects the vertical voltage axis at the same location as V_S.

Characteristics of Practical Voltage Source

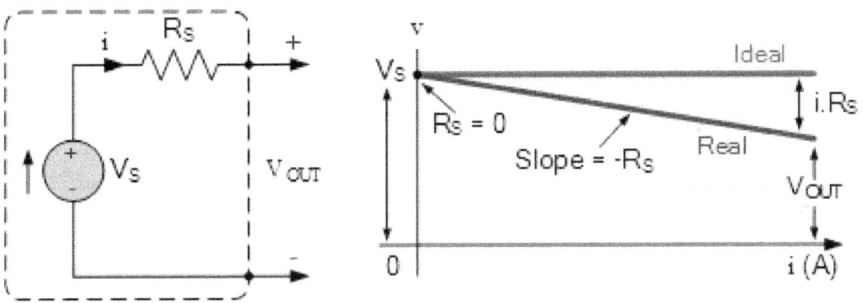

As a result, all real-world, practical voltage sources other than the ideal ones will have an I-V characteristic that is angled down by a factor equal to $i*R_S$, where R_S is the internal source resistance (or impedance). Since the source resistance R_S is typically fairly low, the I-V characteristics of a practical battery provide an extremely near approximation of an ideal voltage source.

Regulating is the process of reducing the slope of the I-V characteristics as the current increases. Since it measures the difference in terminal voltage between an open circuit, or when $I_L = 0$, or no load, and a full load, or when I_L is at maximum, it is a crucial indicator of the practicality of a voltage source (a short-circuit).

Example-2 of Voltage Source

An ideal voltage source is connected in series with an internal resistor to form a battery supply. The battery's terminal voltage and current readings were discovered to be V_{OUT1} = 130V at 10A and V_{OUT2} = 100V at 25A. Determine the optimal voltage source's voltage rating as well as the internal resistance of the source. Draw the I-V characteristics. Let's first define the two voltage and current outputs of the battery supply, denoted as V_{OUT1} and V_{OUT2}, in straightforward "simultaneous equation form."

$$V_{OUT} = V_S - iR_S$$

$$V_{OUT1} = 130 = V_S + 10R_S$$
$$V_{OUT2} = 100 = V_S + 25R_S$$

To determine V_S, we must first multiply V_{OUT1} by five, (5) and V_{OUT2} by two, (2) as shown to make the values of the two currents, I the same for both equations. This is similar to having the voltages and currents in a simultaneous equation form.

$$V_{OUT1} = 130 = V_S + 10R_S \quad \ldots\ldots \times 5$$
$$V_{OUT2} = 100 = V_S + 25R_S \quad \ldots\ldots \times 2$$

$$V_{OUT1} = 650 = 5V_S + 50R_S$$
$$V_{OUT2} = 200 = 2V_S + 50R_S$$

After making the co-efficients for R_S the same by multiplying through with the preceding constants, we now multiply the second equation V_{OUT2} by minus one, (-1) to allow for the subtraction of the two equations, as indicated.

$$V_{OUT1} = 650 = 5V_S + 50R_S$$
$$V_{OUT2} = 200 = 2V_S + 50R_S \quad \ldots\ldots \times -1$$

$$V_{OUT1} = 650 = 5V_S + 50R_S$$
$$V_{OUT2} = -200 = -2V_S - 50R_S$$

Rearrange to give:
$$650 - 200 = (5V_S - 2V_S) + (50R_S - 50R_S)$$
$$450 = 3V_S + 0$$

$$\therefore V_S = \frac{450}{3} = 150V$$

Since we already know that the ideal voltage source, V_S, is 150 volts, we can insert this number into equation V_{OUT1} (or V_{OUT2}, if so desired) and solve for R_S to get the series resistance.

$$V_{OUT1} = 130 = V_S + 10R_S$$

$$V_S = 150\,V$$

$$130 = 150 + 10R_S$$

$$\therefore R_S = \frac{150-130}{10} = 2\,\Omega$$

The battery's internal voltage source is therefore determined to be V_S = 150 volts, and its internal resistance is determined to be R_S = 2 ohms for our straightforward example. The battery's I-V properties are listed as follows:

I-V Characteristics of Battery

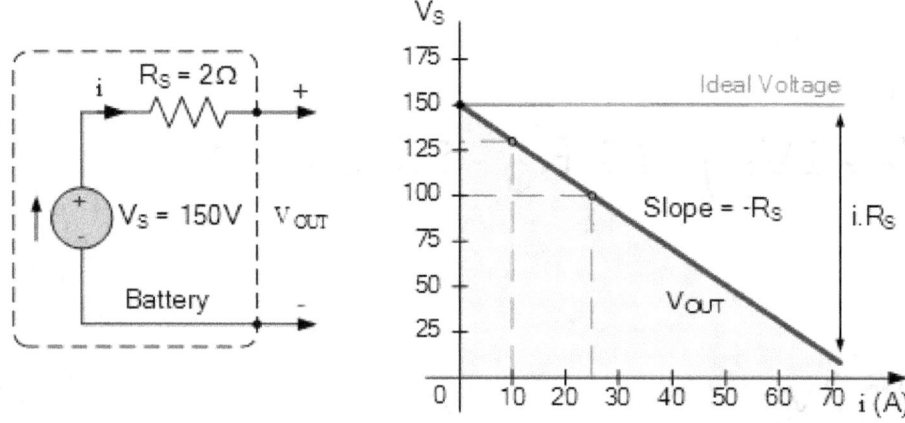

Dependent Voltage Source

A controlled or dependent voltage source, as opposed to an ideal voltage source, which produces a constant voltage across its terminals regardless of what is connected to it, changes its terminal voltage depending on the voltage across or the current through some other element connected to the circuit. Because of this, it can be challenging to specify the value of a dependent voltage source unless you are aware of the precise value of the voltage or current on which it depends.

The behavior of dependent voltage sources is comparable to that of the electrical sources we have previously studied, both real-world and ideal (independent). However, this time, a dependent voltage source can be influenced by an input current or voltage. The term "voltage controlled voltage source," or "VCVS," is used to describe a voltage source that is dependent on a voltage input.

Current Controlled Voltage Sources, or CCVS, are used to describe voltage sources that depend on current input. In order to analyze the input/output properties or the gain of circuit components like operational amplifiers, transistors, and integrated circuits, ideal dependent sources are frequently used. A diamond-shaped symbol, like the one in the illustration, is typically used to denote an ideal voltage dependent source that is either voltage or current controlled.

Symbols of Dependent Voltage Source:

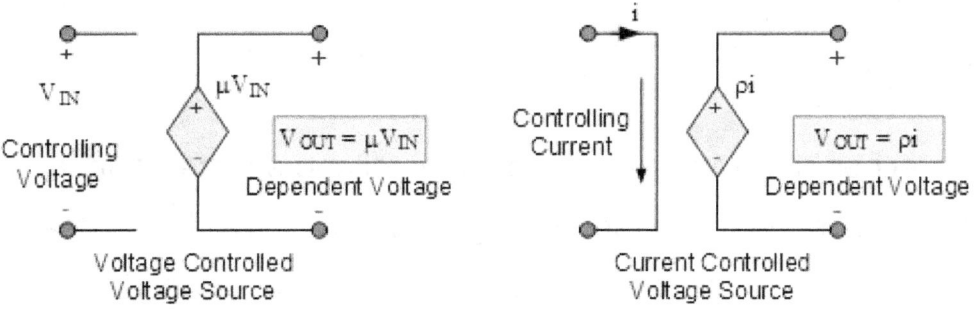

The output voltage of an ideal dependent voltage-controlled voltage source, VCVS, is equal to a multiplying constant (effectively an amplification factor) times the controlling voltage existing elsewhere in the circuit. Because the multiplying constant is, well, a constant, the size of the output voltage, VOUT, is determined by the controlling voltage, VIN.

In other words, the output voltage "depends" on the value of the input voltage, making it a dependent voltage source, thus an ideal transformer can be thought of as a VCVS device, with the turns ratio serving as the amplification factor.

$V_{OUT} = V_{IN}$ is the equation that is used to calculate the VCVS output voltage. Due to the fact that $= V_{OUT}/V_{IN}$ and has the units of volts/volts, it is important to note that the multiplication constant is dimensionless. An ideal dependent current-controlled voltage source, or CCVS, maintains an output voltage equal to some multiplying constant ρ (rho) times a controlling current input generated elsewhere in the connected circuit. The output voltage then begins to "depend" once more on the magnitude of the input current, i.e., becomes a dependent voltage source.

This enables us to describe a current-controlled voltage source as a trans-resistance amplifier with ρ as the multiplication constant, giving us the following equation: $V_{OUT} = I_{IN}$. Since $ρ = V_{OUT}/I_{IN}$, the units of this multiplication constant will be volts/amperes.

Voltage Source in A Brief

A voltage source can either be a regulated dependent voltage source or an ideal independent voltage source. Independent voltage sources provide a steady voltage that is independent of all other circuit components. Batteries, DC generators, or alternators that produce time-varying AC voltage are good independent sources.

Independent voltage sources can be modeled as either a non-ideal or practical voltage source, such as a battery with a resistance connected in series with the circuit to represent the internal resistance of the source, or as an ideal voltage source, ($R_S = 0$), where the output is constant for all load currents. Only ideal voltage sources with the same voltage value can be connected in parallel. The output value will be affected by series-aiding or series-opposing connections.

To assist in the solution of circuit analysis and difficult theorems, voltage sources are short-circuited, causing their voltage to equal zero. It is also important to note that voltage sources can both give and absorb power. A diamond-shaped symbol represents ideal dependent voltage sources that are reliant on and proportionate to an external regulating voltage or current. The multiplication constant µ of a VCVS has no units, whereas the multiplying constant ρ of a CCVS has Ohm's units. A dependent voltage source is useful for modeling electronic or active devices with gain, such as operational amplifiers and transistors.

CHAPTER-12: CURRENT SOURCES

Regardless of the voltage that develops across its terminals, a current source is an active circuit component that can supply a steady current flow to a circuit. A current source, as the name suggests, is a circuit device that consistently maintains a current flow notwithstanding the voltage that develops across its terminals since this voltage is dictated by other circuit elements.

In other words, a perfect constant current source continuously delivers a predetermined amount of current regardless of the impedance it is driving, and as a result, a perfect current source might theoretically offer an endless quantity of energy. Therefore, a current source will likewise have a current rating, such as 3 amperes or 15 amperes, etc., just as a voltage source may be rated, for example, as 5 volts or 10 volts, etc.

Similar to how voltage sources are depicted, ideal continuous current sources do the same, however this time the current source sign is a circle with an arrow within to show which way the current is supposed to flow.

The current will flow out of the positive terminal in a direction that matches the polarity of the corresponding voltage. As illustrated, the letter I is used to signify that the source is current. Ideal Current Source

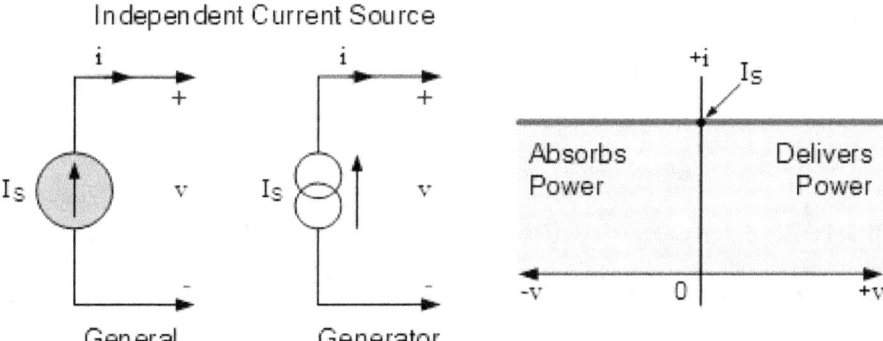

An ideal current source is then referred to as a "constant current source" because it produces an I-V characteristic that is represented by a straight line and offers a constant steady state current regardless of the load attached to it.

The current source can, like voltage sources, be independent (ideal) or dependent (controlled) by a voltage or current present in another part of the circuit. This voltage or current may be constant or time-varying.

For circuits with genuine active parts, ideal independent current sources are often employed to solve circuit theorems and for circuit analysis techniques. A resistor connected in series with a voltage source is the most basic type of current source, producing currents ranging from a few milliamperes to several hundreds of amperes.

A current source with a zero value is an open circuit since R = 0. A two-terminal component that permits the current flow shown by the arrow is what is meant by the term "current source." A current source then has a value, I measured in amperes (A), which are commonly shortened to amps.

Ohm's law, which assumes that both the voltage and current variables in a network will have fixed values, describes the physical relationship between a current source and these variables. The magnitude and polarity of the voltage produced by an ideal current source as a function of the current may be difficult to determine, especially if the associated circuit contains additional voltage or current sources.

The voltage across the current source may thus be unknown unless the power supplied by the current source is specified as P = V*I. In this case, we may only be aware of the current supplied by the current source. It will be simpler to determine the polarity of the voltage across the source if the circuit's current source is the only source present. The terminal voltage will be determined by the network that each source is linked to if there are several sources, though.

Connecting Current Sources Together

Ideal current sources can also be connected together to enhance (or decrease) the available current, just like voltage sources can. However, there exist guidelines for connecting two or more independent current sources, whether in series or parallel, with distinct values.

Current Source in Parallel

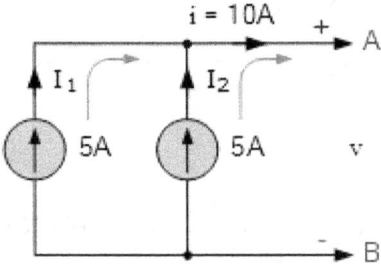

When two or more current sources are connected in parallel, they become one current source, whose total current output is calculated by adding the individual source currents algebraically. As seen in this example, $I_T = I_1 + I_2$ combines two 5 amp current sources to produce 10 amps. Different current sources of values could be connected in parallel.

For instance, since the arrows indicating the current source point in the same direction, one of 5 amps and one of 3 amps combined would result in a single current source of 8 amperes. The two currents' link is then described as parallel-aiding when they combine.

While not best practice for circuit analysis, parallel-opposing connections use current sources that are connected in opposite directions to form a single current source whose value is the algebraic subtraction of the individual sources.

Parallel Opposing Current Sources

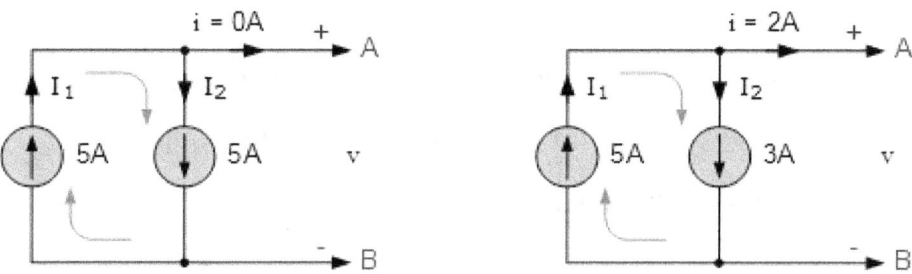

In this case, the two current sources provide a closed-loop path for a circulating current that complies with Kirchoff's Current Law, KCL, as a result of being connected in opposing directions (shown by their arrows). Therefore, two current sources that are 5 amps each would produce zero output since 5A -5A = 0A. The output will also be the subtracted value with the smaller current subtracted from the bigger current if the two currents are different values, such as 5A and 3A.

Hence, the IT is 5 – 3 = 2A. The ability to create parallel-aiding or parallel-opposing current sources by connecting ideal current sources in parallel has been demonstrated. Connecting perfect current sources in series combinations is not permitted or recommended for circuit study.

Current Sources in Series

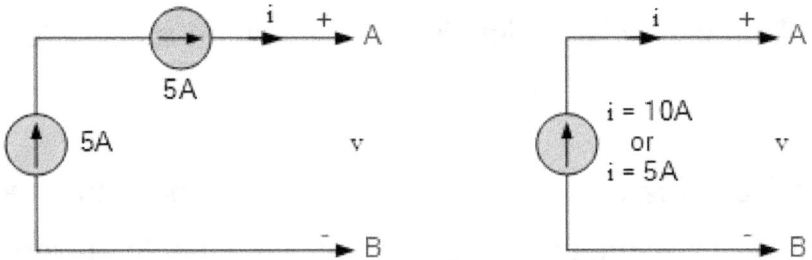

Both current sources with the same value and those with different values cannot be connected in series. What is the resultant current value in this example when two current sources with a capacity of 5 amps each are connected in series?

Does it equal the sum of the two sources, which is 10 amps, or one source at 5 amps? This introduces an unknown element into circuit analysis from series coupled current sources, which is undesirable.

The possibility that they might not produce the same current in the same direction is another reason why series coupled sources are not permitted for circuit analysis techniques. Current sources that are optimal don't have series-aiding or series-opposing currents.

Example-1 of Current Source

To supply a connected load of 20 ohms, two current sources with respective outputs of 250 milliamps and 150 milliamps are connected in parallel. Calculate the power dissipated as well as the voltage drop across the load. Draw the circuit.

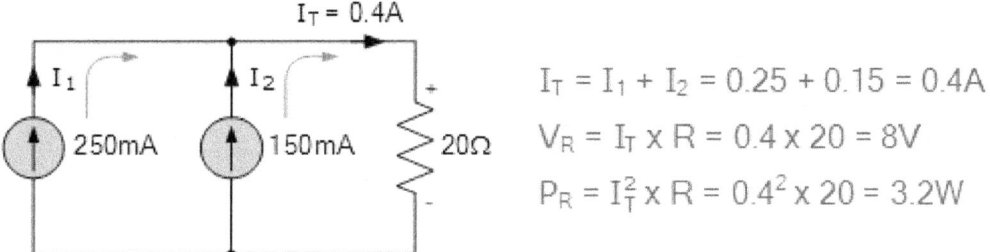

$I_T = I_1 + I_2 = 0.25 + 0.15 = 0.4A$

$V_R = I_T \times R = 0.4 \times 20 = 8V$

$P_R = I_T^2 \times R = 0.4^2 \times 20 = 3.2W$

Then, I_T = 0.4A or 400mA, V_R = 8V, and P_R = 3.2W

Practical Current Source

It has been observed that an ideal constant current source can give the same amount of current forever regardless of the voltage across its terminals, making it a source that can be used independently. As a result, the current source has an infinite internal resistance (R = ∞).

This theory works well for circuit analysis techniques, but in practice, practical current sources always contain an internal resistance, no matter how huge (typically in the mega-ohms region), causing the generated source to vary slightly with the load. An ideal source with an internal resistance attached across it can be used to represent an actual or non-ideal current source. The internal resistance (R_P) has the same result as a parallel (shunt) resistance connection to the current source, as indicated. Keep in mind that parallel circuit elements all have the same voltage drop across them.

Ideal and Practical Current Source

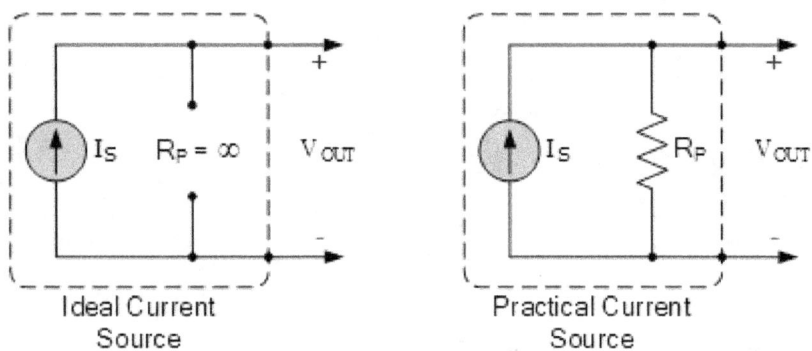

Ideal Current Source   Practical Current Source

A practical current source may resemble a Norton's equivalent circuit, as Norton's theorem claims that "any linear dc network can be substituted by an equivalent circuit consisting of a constant-current source, IS in parallel with a resistor, R_P". It should be noted here that if the parallel resistance is extremely low, $R_P = 0$, the current source is short-circuited. The current source can be treated as perfect when the parallel resistance is very high or infinite, $R_P \approx \infty$.

On the I-V characteristic, as previously mentioned, the ideal current source displays a horizontal line.

The characteristic of this practical source is not flat and horizontal, but will reduce as the current is now splitting into two parts, with one part of the current flowing into the parallel resistance, R_P, and the other part of the current flowing directly to the output terminals. This is because practical current sources have an internal source resistance, which absorbs some of the current.

According to Ohm's law, a voltage drop occurs across a resistance (R) when a current I passes through it. This voltage drop's value will be provided as $i*R_P$. When there is no load connected to the resistor, the voltage drop across it will equal V_{OUT}. We should keep in mind that for a perfect source current, R_P is infinite since there is no internal resistance, and as a result, the terminal voltage will be 0 because there is no voltage drop.

Kirchoff's current law, KCL, states that $I_{OUT} = I_S - V_S/R_P$ is the total current through the loop. The I-V characteristics of the output current can be determined by plotting this equation. When the source is ideal as indicated, it is represented as a straight line with a slope of $-R_P$ that meets the vertical voltage axis at the same location as I_S.

Characteristics of Practical Current Source

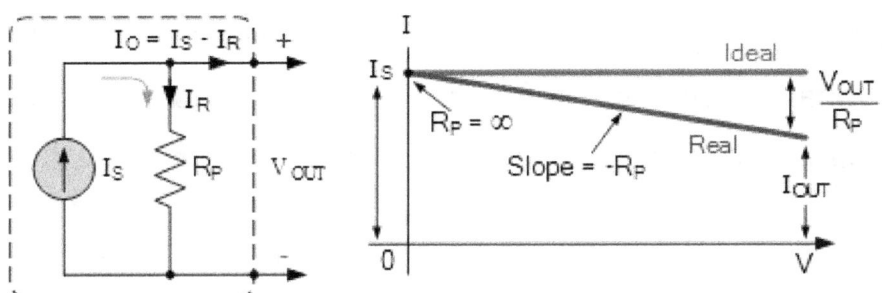

All ideal current sources will therefore have a straight I-V characteristic, whereas real-world, practical current sources will have an I-V characteristic that is slightly inclined downward by a factor equal to V_{OUT}/R_P, where R_P is the internal source resistance.

Example-2 of Current Source

A 3A ideal current source with a 500 Ohm internal resistance serves as a useful current source. Calculate the internal resistor's no-load power consumption and the current source's open-circuit terminal voltage with no load applied.

1. No-load values:

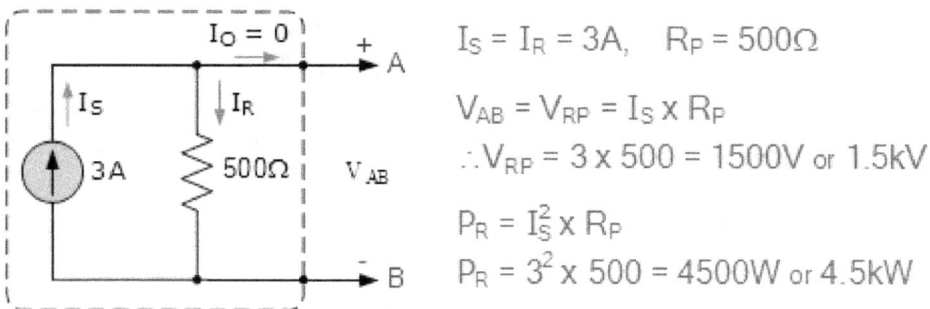

$I_S = I_R = 3A, \quad R_P = 500\Omega$

$V_{AB} = V_{RP} = I_S \times R_P$

$\therefore V_{RP} = 3 \times 500 = 1500V$ or $1.5kV$

$P_R = I_S^2 \times R_P$

$P_R = 3^2 \times 500 = 4500W$ or $4.5kW$

After that, 1500 volts are calculated as the open circuit voltage (VAB) across the internal source resistance and terminals A and B.

Part 2: Determine the current through each resistance, the power absorbed by each resistance, and the voltage drop across the load resistor if a 250 Ohm load resistor is attached to the terminals of the same realistic current source. Draw the circuit.

2. Data provided with a load attached:

$R_L = 250\Omega$, $R_P = 500\Omega$, and $I_S = 3A$.

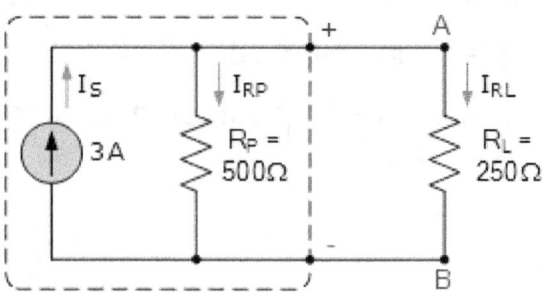

2a. To determine the currents in each resistive branch, current-division rule is applied.

$$I_{RP} = \frac{R_L}{R_P + R_L} \times I_S = \frac{250}{500 + 250} \times 3 = 1A$$

$$I_{RL} = \frac{R_P}{R_L + R_P} \times I_S = \frac{500}{250 + 500} \times 3 = 2A$$

$$\therefore I_{RP} + I_{RL} = 1 + 2 = 3A = I_S$$

2b. The power absorbed by each resistor is as follows:

$$P_{RP} = I_{RP}^2 \times R_P = 1^2 \times 500 = 500W$$

$$P_{RL} = I_{RL}^2 \times R_L = 2^2 \times 250 = 1000W$$

2c. Voltage drop across the load resistor, R_L is as follows:

$$V_{AB} = I_S \times R_T$$

$$R_T = \frac{R_P \times R_L}{R_P + R_L} = \frac{500 \times 250}{500 + 250} = 166.7\Omega$$

$$\therefore V_{AB} = 3 \times 166.7 = 500V$$

It has been observed that an open-circuited practical current source can have very high terminal voltages because it can generate whatever voltage required—in this case, 1500 volts—to give the required current.

The source tries to give the rated current, but theoretically this terminal voltage may be limitless. As the current now has a somewhere to go and for a constant current source, the terminal voltage is precisely proportional to the load resistance, connecting a load across the source's terminals will lower the voltage, in this case 500 volts.

The total internal resistance (or impedance) of a set of non-ideal current sources will be determined by adding all of the sources together in parallel, just as it is with parallel resistors.

Dependent Current Source

As a result, the ideal current source will generate whatever voltage is required to maintain the desired current. As we now know, an ideal current source generates a specific amount of current absolutely irrespective of the voltage across it. As a result, it becomes wholly independent of the circuit to which it is linked, earning the moniker "perfect independent current source."

On the other hand, a regulated or dependent current source modifies the available current in response to the voltage across or the current through another component connected to the circuit. In other words, another voltage or current controls the output of a dependent current source.

Dependent current sources function similarly to the previous current sources in that they are both ideal (independent) and practical. This time, though, a dependent current source can be controlled by an input voltage or current.

A Voltage Controlled Current Source, or VCCS, is a current source that is controlled by a voltage input. A current source that is controlled by a current input is known as a Current Controlled Current Source, or CCCS. An ideal current dependent source, either voltage or current controlled, is generally shown by a diamond-shaped symbol with an arrow indicating the current's direction, I as illustrated.

Symbols of Dependent Current Source

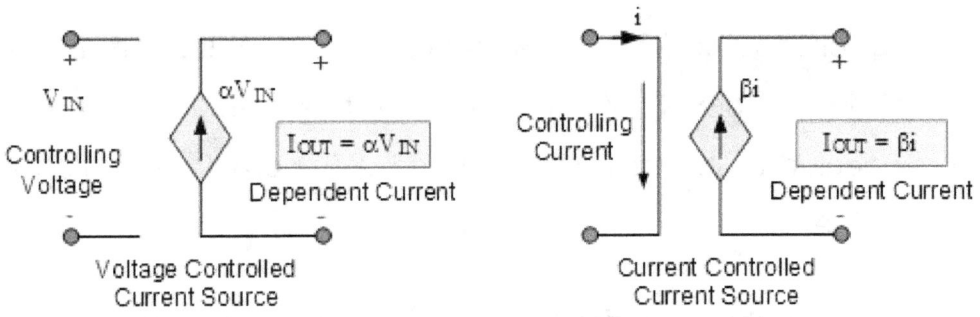

Voltage Controlled Current Source

Current Controlled Current Source

The output current, I_{OUT}, of an ideal dependent voltage-controlled current source (VCCS) is proportional to the controlling input voltage, V_{IN}. In other words, it is a dependent current source since the output current "depends" on the magnitude of the input voltage.

VCCS output current equation can be written as: $I_{OUT} = \alpha V_{IN}$. Since $\alpha = I_{OUT}/V_{IN}$ and hence has amperes/volt as its units, this multiplication constant, α (alpha) has the SI units of mhos, ℧ or (an inverted Ohms sign).

A dependent current-controlled current source (CCCS) in its optimal state maintains an output current that is proportionate to the controlling input current. Once more becoming a dependent current source, the output current then "depends" on the magnitude of the incoming current.

Since a controlling current, I_{IN} finds the magnitude of the output current, I_{OUT} times the magnification constant β (beta), the output current for a CCCS element is obtained by the following equation: $I_{OUT} = \beta I_{IN}$. It should be noted here that the multiplying constant β is a dimensionless scaling factor as $\beta = I_{OUT}/I_{IN}$. As a result, its units would be amperes/amperes.

Current Source in A Brief

We know that an ideal current source ($R = \infty$) is an active element that produces a constant current that is completely independent of the voltage across it because of the load connected to it, which results in an I-V characteristic that is represented by a straight line.

For the purposes of circuit analysis methodologies, ideal independent current sources can be coupled together in parallel in either parallel-aiding or parallel-opposing topologies, but not in series.

Current sources are converted to open-circuited sources to make their current equal zero in order to solve circuit analysis and theorems. Also keep in mind that current sources have the ability to produce or consume electricity.

Current sources that are not ideal or practical can be modeled as an equivalent ideal current source and an internal parallel (shunt) connected resistance that is not infinite but has a very high value as $R \approx \infty$, resulting in an I-V characteristic that is not straight but slopes downward as the load decreases.

Additionally, we have shown that current sources might be dependent or independent in this case. A source is said to be dependent if its value depends on another circuit variable. Dependent current sources include voltage-controlled current sources (VCCS) and current-controlled current sources (CCCS).

It is possible to construct constant current sources with extremely high internal resistances utilizing bipolar transistors, diodes, zeners, FETs, as well as a combination of these solid-state components. These sources are used in many electronic circuits and analyses.

CHAPTER-13: KIRCHHOFF'S CURRENT LAW

The first law of Kirchhoff that addresses the conservation of charge entering and exiting a junction is known as the Kirchhoff's Current Law (KCL). Using specific principles or rules that enable us to express these currents as an equation, we can determine the size or magnitude of the electrical current flowing around an electrical or electronic circuit. We shall examine Kirchhoff's current law as we are working with circuit currents and the network equations employed are those according to Kirchhoff's laws (KCL).

One of the fundamental laws utilized in circuit analysis is Gustav Kirchhoff's Current Law. According to his current theory, the total current entering a circuit junction is exactly identical to the total current exiting the same junction on a parallel line. This is because, since no charge is lost, it has nowhere else to go.

In other terms, Σ IIN = Σ IOUT means that the algebraic sum of ALL currents entering and exiting a junction must be equal to zero. Since there is no current loss surrounding the junction, Kirchhoff's theory is also known as the "Conservation of Charge." Let's see a straightforward illustration of Kirchhoff's current law (KCL) in action at a single junction.

A Single Junction

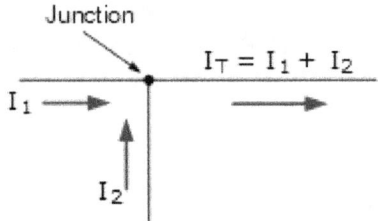

In this straightforward example of a single junction, the current I_T is the algebraic sum of the two currents I1 and I2, which are entering the junction. I_1 and I_2 add together to form I_T. The right way to write this is also as the algebraic total, which is: $I_T - (I_1 + I_2) = 0$. Take note of this.

Therefore, if I_1 is equal to 3 amperes and I_2 is equal to 2 amperes, the total current, I_T exiting the junction will be 3 + 2 = 5 amperes. We can apply this fundamental law to any number of junctions or nodes because the sum of the currents coming in and going out will always be the same.

Additionally, the derived equations would still hold true for I_1 or I_2 if we switched the currents' directions. Since $I_2 = I_T - I_1 = 5 - 3 = 2$ amps and $I_1 = I_T - I_2 = 5 - 2 = 3$ amps, respectively. Thus, the currents entering the junction can be thought of as positive (+), while the ones leaving the junction can be thought of as negative (-).

The Kirchhoff's Junction Rule, also known as Kirchhoff's Current Law or Kirchhoff's Current Law, is based on the observation that the mathematical sum of the currents entering or leaving the junction in any direction would always equal zero (KCL).

Resistors in Parallel

Consider how we might apply Kirchhoff's current rule to parallel resistors, regardless of whether the resistances in those branches are equal or unequal. Consider the circuit diagram below:

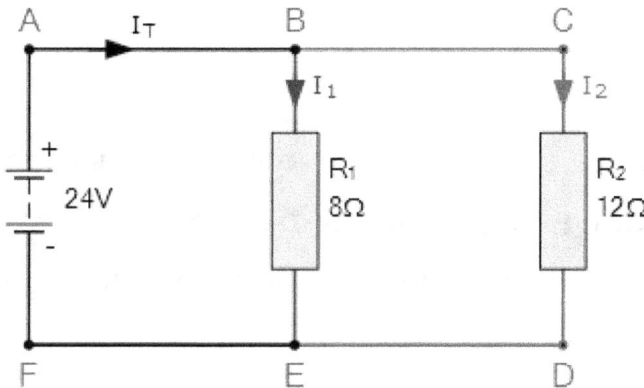

In this straightforward parallel resistor example, the current has two separate junctions. Node B is where junction one happens, and node E is where junction two happens. Kirchhoff's Junction Rule may be applied to the electrical currents going into and out of these two separate junctions, allowing us to compare their electrical currents.

All of the current initially leaves the 24 volt source, travels to point A, and then enters node B. Node B is a junction because the current can now flow in two different directions, with some of it passing via resistor R1 and downward, and the other half passing through resistor R2 and upward through node C.

Keep in mind that the currents entering and leaving a node point are often referred to as branch currents. Since I = V/R, we can use Ohm's Law to calculate the individual branch currents across each resistor. Branch B to E's current path is through resistor R_1.

$$I_{B-E} = I_1 = \frac{V}{R_1} = \frac{24}{8} = 3A$$

R_2 acts as a resistor for the current branch from C to D.

$$I_{C-D} = I_2 = \frac{V}{R_2} = \frac{24}{12} = 2A$$

Kirchhoff's current law states that the whole sum of currents entering a junction must equal the total sum of currents leaving the junction. In our straightforward example above, there is one current, I_T, entering the junction at node B and two currents, I_1 and I_2, exiting the junction.

The sum of the currents entering the junction at node B must be equal to 3 + 2 = 5 amps as we can now calculate that the currents exiting the junction at node B are I_1 equals 3 amps and I_2 equals 2 amps. Therefore, $\Sigma I_N = I_T$ = 5 amps.

Since nodes B and E in our example have two separate junctions, we can confirm this value for I_T because the two currents recombine at node E. Kirchhoff's junction rule requires that the total current flowing into point F match the total current leaving the junction at node E in order for it to be valid.

The sum of the currents entering point F is consequently 3 + 2 = 5 amperes because the two currents entering junction E are 3 amps and 2 amps, respectively. As a result, Kirchhoff's current law is valid since $\Sigma I_N = I_T = 5$ amperes, which is also the same value as the current exiting point A.

Using KCL in more complex circuits

To determine the currents circulating within increasingly complicated circuits, we can apply Kirchhoff's Current Law (KCL). According to KCL, the algebraic sum of all the currents at a node (junction point) equals zero, therefore finding the currents entering and exiting the node is a straightforward process. Take a look at the circuit below.

Example-1 of Kirchhoff's Current Law

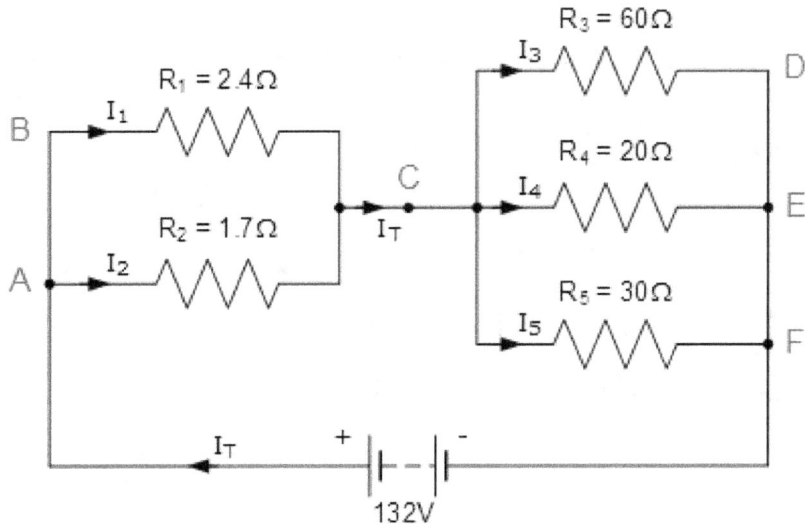

There are four unique junctions in this example where current can split or combine at nodes A, C, E, and node F. Supply current I_T splits at node A, passing through resistors R_1 and R_2, recombining at node C, splitting once more through resistors R_3, R_4, and R_5, and ultimately splitting once more through resistors R_3, R_4, and R_5.

However, we must first determine the total current (I_T) of the circuit in order to calculate the individual currents flowing through each resistor branch. We may compute the circuit resistances using the formula I = V/R, where V is the known value of 132 volts.

Circuit Resistance R_{AC}

$$\frac{1}{R_{(AC)}} = \frac{1}{R_1} + \frac{1}{R_2} = \frac{1}{2.4} + \frac{1}{1.7}$$

$$\frac{1}{R_{(AC)}} = 1 \quad \therefore R_{(AC)} = 1\Omega$$

In this way, the equivalent circuit resistance between nodes A and C is calculated as 1 Ohm.

Circuit Resistance R_{CF}

$$\frac{1}{R_{(CF)}} = \frac{1}{R_3} + \frac{1}{R_4} + \frac{1}{R_5} = \frac{1}{60} + \frac{1}{20} + \frac{1}{30}$$

$$\frac{1}{R_{(CF)}} = 0.1 \quad \therefore R_{(CF)} = 10\Omega$$

In this way, the equivalent circuit resistance between nodes C and F is calculated as 10 Ohms. So, the total circuit current, I_T will be as follows:

$$R_T = R_{(AC)} + R_{(CF)} = 1 + 10 = 11\,\Omega$$

$$I_T = \frac{V}{R_T} = \frac{132}{11} = 12\ \text{Amperes}$$

Providing us an equivalent circuit of:

Kirchhoff's Current Law Equivalent Circuit

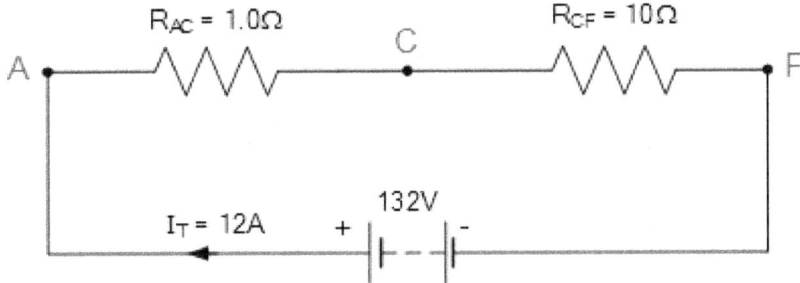

As a result, V = 132V, R_{AC} = 1Ω, R_{CF} = 10Ω's and I_T = 12A.

Once the supply current and equivalent parallel resistances have been established, we can use Kirchhoff's junction rule to calculate the individual branch currents and confirm.

$$V_{AC} = I_T \times R_{AC} = 12 \times 1 = 12 \text{ Volts}$$

$$V_{CF} = I_T \times R_{CF} = 12 \times 10 = 120 \text{ Volts}$$

$$I_1 = \frac{V_{AC}}{R_1} = \frac{12}{2.4} = 5 \text{ Amps}$$

$$I_2 = \frac{V_{AC}}{R_2} = \frac{12}{1.7} = 7 \text{ Amps}$$

$$I_3 = \frac{V_{CF}}{R_3} = \frac{120}{60} = 2 \text{ Amps}$$

$$I_4 = \frac{V_{CF}}{R_4} = \frac{120}{20} = 6 \text{ Amps}$$

$$I_5 = \frac{V_{CF}}{R_5} = \frac{120}{30} = 4 \text{ Amps}$$

In this way, I_1 = 5A, I_2 = 7A, I_3 = 2A, I_4 = 6A, and I_5 = 4A. By utilizing node C as our reference point and computing the currents arriving and leaving the junction as follows, we can verify that Kirchoff's current law is valid throughout the circuit:

At node C $\sum I_{IN} = \sum I_{OUT}$

$I_T = I_1 + I_2 = I_3 + I_4 + I_5$

$\therefore 12 = (5 + 7) = (2 + 6 + 4)$

The fact that the currents entering and leaving the junction are positive and zero, respectively, allows us to confirm the validity of Kirchhoff's Current Law. Accordingly, the algebraic sum of $I_1 + I_2 - I_3 - I_4 - I_5$ is $5 + 7 - 2 - 6 - 4 = 0$.

The algebraic total of the currents at a junction point in a circuit network is always zero, according to Kirchhoff's current law (KCL), which is accurate and valid in this example.

Example-2 of Kirchhoff's Current Law

Using only Kirchhoff's Current Law, determine the currents that are flowing through the following circuit.

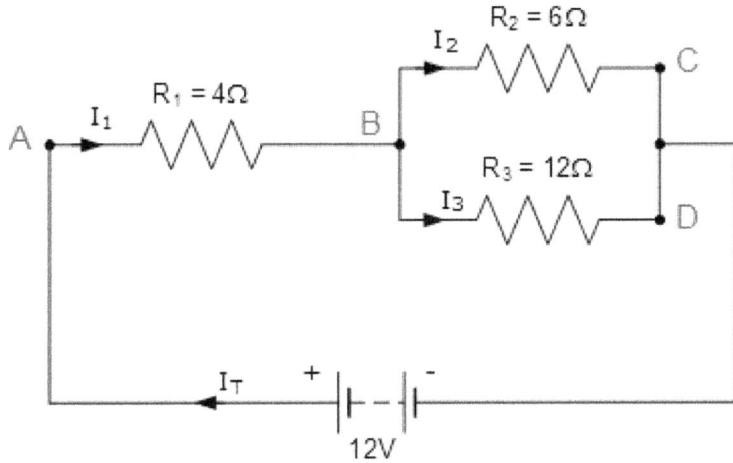

I_T is the circuit's overall current, which is driven by the 12V supply voltage. I_1 is equal to I_T at point A, causing a voltage drop of I_1*R across resistor R_1. The circuit has two independent loops, three nodes (B, C, and D), and two branches. As a result, the I*R voltage drops around the two loops will be as follows:

Loop ABC $\Rightarrow$ $12 = 4I_1 + 6I_2$

Loop ABD $\Rightarrow$ $12 = 4I_1 + 12I_3$

We may therefore substitute current I_1 for $(I_2 + I_3)$ in both of the following loop equations and then simplify because Kirchhoff's current rule specifies that at node B, $I_1 = I_2 + I_3$.

Kirchhoff's Loop Equations

Loop (ABC)   Loop (ABD)

$12 = 4I_1 + 6I_2$ $12 = 4I_1 + 12I_3$

$12 = 4(I_2 + I_3) + 6I_2$ $12 = 4(I_2 + I_3) + 12I_3$

$12 = 4I_2 + 4I_3 + 6I_2$ $12 = 4I_2 + 4I_3 + 12I_3$

$12 = 10I_2 + 4I_3$ $12 = 4I_2 + 16I_3$

Now, we have two equations that are simultaneous and relate to the currents flowing through the circuit.

Eq. No 1: $12 = 10I_2 + 4I_3$

Eq. No 2: $12 = 4I_2 + 16I_3$

We can reduce both equations to give us the values of I_2 and I_3 by multiplying the first equation (Loop ABC) by 4 and subtracting Loop ABD from Loop ABC.

Eq. No 1: $12 = 10I_2 + 4I_3$ (x4) $\Rightarrow$ $48 = 40I_2 + 16I_3$

Eq. No 2: $12 = 4I_2 + 16I_3$ (x1) $\Rightarrow$ $12 = 4I_2 + 16I_3$

Eq. No 1 − Eq. No 2 $\Rightarrow$ $36 = 36I_2 + 0$

The value of I_2 is given as 1.0 Amps by substituting I_2 for I_3. By multiplying the first equation (Loop ABC) by 4 and the second equation (Loop ABD) by 10, we can now follow the same steps to determine the value of I_3. Once more, we may decrease both equations to give us the values of I_2 and I_3 by subtracting Loop ABC from Loop ABD.

Eq. No 1 : $12 = 10I_2 + 4I_3$ (x4) $\Rightarrow$ $48 = 40I_2 + 16I_3$

Eq. No 2 : $12 = 4I_2 + 16I_3$ (x10) $\Rightarrow$ $120 = 40I_2 + 160I_3$

Eq. No 2 – Eq. No 1 $\Rightarrow$ $72 = 0 + 144I_3$

In this way, we can get the value of I_3 as 0.5 Amps by substituting of I_3 in terms of I_2. According to Kirchhoff's junction rule: $I_1 = I_2 + I_3$. Supply current flowing through resistor R_1 will be: 1.0 + 0.5 = 1.5 Amps

We may calculate the I*R voltage drops across the devices and at the various locations (nodes) throughout the circuit using the information that $I_1 = I_T = 1.5$ Amps, $I_2 = 1.0$ Amps, and $I_3 = 0.5$ Amps.

Although Kirchhoff's Current Law can be used to solve more complex circuits when Ohm's Law cannot be used directly, we could have solved the circuit in Example 2 just as easily and quickly with that law alone.

CHAPTER-14: KIRCHHOFF'S VOLTAGE LAW

Kirchhoff's Voltage Law (KVL) is concerned with the energy conservation in a closed circuit channel. The second of Gustav Kirchhoff's essential rules that we can apply to circuit analysis is the voltage law.

According to Kirchhoff's voltage law, the algebraic total of all the voltages around any closed loop in a circuit is equal to zero for a closed loop series path. This is so that no energy is lost because a circuit loop has a closed conducting line. To put it another way, $\Sigma V = 0$ means that the algebraic sum of ALL possible differences around the loop must equal zero.

It should be noted that the phrase "algebraic sum" refers to accounting for the polarities, signs, and voltage decreases around the loop. The Conservation of Energy, a Kirchhoffian principle, states that if you move through a closed circuit or loop, you will eventually return to your starting point and, consequently, to the same beginning potential with no loss of voltage. So, any voltage losses experienced while completing the loop must equal any voltage sources encountered along the route.

Therefore, it is crucial to pay close attention to the algebraic signs, (+ and -), of the voltage drops across elements and the emfs of sources when applying Kirchhoff's voltage law to a particular circuit element. Otherwise, our calculations may be incorrect. Kirchhoff's voltage law (KVL) will be discussed later, but first, let's examine how much voltage is lost across a single element, such a resistor.

A Single Electrical Circuit Component

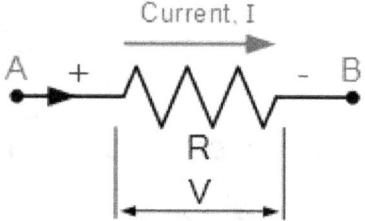

Here, let's consider that the direction of the current, I, which is typical current flow, is the same as the direction of the flow of positive charge. Here, the resistor's current is flowing from a positive terminal to a negative terminal, or from point A to point B. As a result, because we are moving in the same direction as current flow, there will be a reduction in potential across the resistive element, resulting in a -IR voltage drop across it.

The resistive element would experience a rise in potential as we move from a - potential to a + potential, resulting in a +I*R voltage drop, if the current flowed in the opposite direction from point B to point A. Thus, in order to accurately apply Kirchhoff's voltage law to a circuit, we must first know the polarity, and as we can see, the sign of the voltage drop across the resistive element depends on the polarity of the current passing through it. As a general rule, you will gain potential as you go in the direction of an emf source and lose potential as you move in the same direction as current across an element.

Either clockwise or counterclockwise current flow can be chosen as the direction of current flow around a closed circuit. The solution will still be correct and valid if the chosen direction is different from the actual direction of current flow, but the algebraic answer will have a minus sign. Let's examine a single circuit loop to check if Kirchhoff's Voltage Law is valid to gain a better understanding of this concept.

A Single Circuit Loop

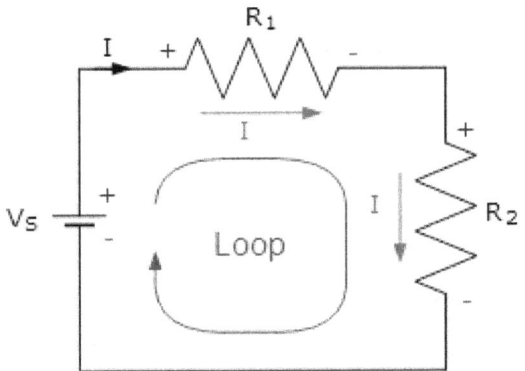

According to Kirchhoff's voltage law, any loop's algebraic sum of potential differences must equal zero, with $\Sigma V = 0$. Since the two resistors, R_1 and R_2, are connected in series, they are both a part of the same loop and require the same amount of current to flow through them. Therefore, the voltage drop across resistor, R_1, and the voltage drop across resistor, R_2, given by KVL, are equal to $I*R_1$ and $I*R_2$, respectively.

$$V_S + (-IR_1) + (-IR_2) = 0$$

$$\therefore V_S = IR_1 + IR_2$$

$$V_S = I(R_1 + R_2)$$

$$V_S = IR_T$$

Where: $R_T = R_1 + R_2$

Kirchhoff's Voltage Law can be used to calculate the equivalent or total resistance in a series circuit for a single closed loop. From there, we can use an extension of this method to calculate the voltage drops around the loop.

$$R_T = R_1 + R_2$$

$$I = \frac{V_S}{R_T} = \frac{V_S}{R_1 + R_2}$$

$$V_{R1} = IR_1 = V_S \left(\frac{R_1}{R_1 + R_2} \right)$$

$$V_{R2} = IR_2 = V_S \left(\frac{R_2}{R_1 + R_2} \right)$$

Example-1 of Kirchhoff's Voltage Law

Three resistors, each with a value of 10 ohms, 20 ohms, and 30 ohms, are linked in series across a 12 volt power source. Make the following calculations to ensure Kirchhoff's voltage law, KVL, is valid: a) the total resistance; b) the circuit current; c) the current flowing through each resistor; d) the voltage drop across each resistor.

a) Total Resistance (R_T)

$R_T = R_1 + R_2 + R_3 = 10Ω + 20Ω + 30Ω = 60Ω$

Therefore, the total circuit resistance R_T is equal to 60Ω

b) Circuit Current (I)

$$I = \frac{V_S}{R_T} = \frac{12}{60} = 0.2A$$

In this way, the total circuit current I is equal to 0.2 amperes or 200mA

c) Current Through Each Resistor

Since the resistors are connected in series, they are all a part of the same circuit and receive the same amount of current. In this way:

$I_{R1} = I_{R2} = I_{R3} = I_{SERIES} = 0.2$ amperes

d) Voltage Drop Across Each Resistor

$V_{R1} = I \times R_1 = 0.2 \times 10 = 2$ volts

$V_{R2} = I \times R_2 = 0.2 \times 20 = 4$ volts

$V_{R3} = I \times R_3 = 0.2 \times 30 = 6$ volts

e) Verify Kirchhoff's Voltage Law

$$V_S + (-IR_1) + (-IR_2) + (-IR_3) = 0$$

$$12 + (-0.2 \times 10) + (-0.2 \times 20) + (-0.2 \times 30) = 0$$

$$12 + (-2) + (-4) + (-6) = 0$$

$$\therefore 12 - 2 - 4 - 6 = 0$$

Kirchhoff's voltage law is therefore valid as long as the sum of the individual voltage drops along the closed loop.

Kirchhoff's Circuit Loop

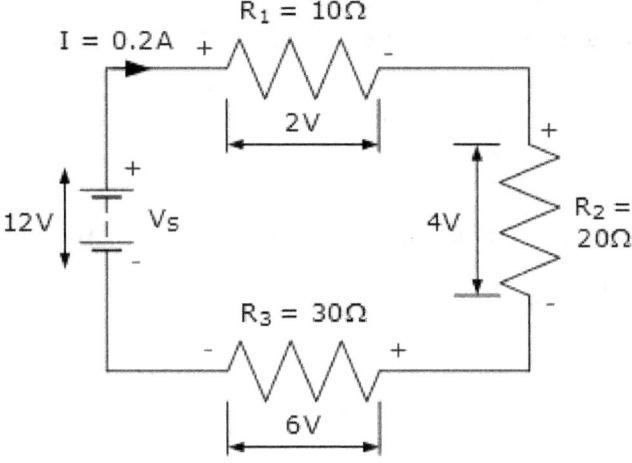

Kirchhoff's voltage law, or KVL, is Kirchhoff's second law which defines that the algebraic total of all voltage drops experienced while traveling around a closed circuit starting at one fixed point and ending at the same place, while accounting for polarity, is always zero. As a result, ΣV = 0.

The Kirchhoff's second law's underlying theory, also known as the law of conservation of voltage, is very helpful for us when working with series circuits because these circuits also function as voltage dividers, which is a critical function for many series circuits.

CHAPTER-15: VOLTAGE DIVIDERS

Although all of the components in a series circuit share the same current, voltage divider circuits can produce several voltage levels from a single voltage source. Using a common source voltage, voltage divider circuits can produce several voltage levels. This common supply could be across a dual supply, such as ±5V, or ±12V, or it could be a single supply that is either positive or negative, such as +5V, +12V, -5V or -12V, etc., with regard to a common point or ground, which is typically 0V.

Since the "Volt," the voltage unit, measures the magnitude of the potential difference between two points, voltage dividers are sometimes referred to as potential dividers. A voltage or potential divider is a straightforward passive circuit that makes use of the voltage drop that occurs when components are connected in series.

The simplest straightforward example of a voltage divider is a potentiometer, which is a variable resistor with a sliding contact. We can apply a voltage across its terminals and generate an output voltage proportional to the mechanical position of its sliding contact.

However, since they are two-terminal devices that can be connected in series, individual resistors, capacitors, and inductors can also be used to form voltage dividers.

Resistive Voltage Divider

The most basic type of a passive voltage divider network is two resistors connected in series. This fundamental combination enables us to utilize the Voltage Divider Rule to determine the voltage decreases across each series resistor.

Resistive Voltage Divider Circuit

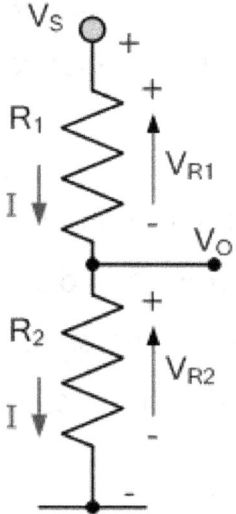

Two resistors, R_1 and R_2, are wired in series to form the circuit in this instance. As there is no other place for the electric current to travel in the circuit due to the two resistors are connected in series, it follows that each resistive component must experience the same amount of electric current flow. It provides each resistive element an I*R voltage drop.

When a supply or source voltage, V_S, is applied across this series combination, Kirchhoff's Voltage Law (KVL) and Ohm's Law can be used to determine the voltage dropped across each resistor in terms of the shared current, I, that is flowing through them. In order to find the current (I) flowing through the series network, we can solve for:

$$V_S = V_{R1} + V_{R2} \quad (KVL)$$

$$V_{R1} = I \times R_1 \quad \text{and} \quad V_{R2} = I \times R_2$$

$$\text{Then}: V_S = I \times R_1 + I \times R_2$$

$$\therefore V_S = I(R_1 + R_2)$$

$$\text{So}: I = \frac{V_S}{(R_1 + R_2)}$$

Following Ohm's Law, the current flowing across the series network is just I = V/R. Given that the current flows through both resistors equally ($I_{R1} = I_{R2}$), we can determine the voltage dropped across resistor R_2 in the series circuit shown above as follows:

$$I_{R2} = \frac{V_{R2}}{R_2} = \frac{V_S}{(R_1 + R_2)}$$

$$\therefore V_{R2} = V_S \left(\frac{R_2}{R_1 + R_2} \right)$$

Similarly, for resistor R_1 as being:

$$I_{R1} = \frac{V_{R1}}{R_1} = \frac{V_S}{(R_1 + R_2)}$$

$$\therefore V_{R1} = V_S \left(\frac{R_1}{R_1 + R_2} \right)$$

Example-1 of Voltage Divider

When the supply voltage across the series combination is 12 volts dc, how much current will flow through a 20Ω resistor connected in series with a 40Ω resistor? Calculate the voltage drop produced by each resistor as well.

$$R_T = R_1 + R_2 = 20 + 40 = 60\Omega$$

$$I = \frac{V_S}{R_T} = \frac{12}{60} = 0.2 \text{ [Amps, A] or } 200mA$$

$$V_{R1} = I \times R_1 = V_S\left(\frac{R_1}{R_1+R_2}\right) = 12\left(\frac{20}{20+40}\right) = 4\text{volts}$$

$$V_{R2} = I \times R_2 = V_S\left(\frac{R_2}{R_1+R_2}\right) = 12\left(\frac{40}{20+40}\right) = 8\text{volts}$$

In proportion to its resistive value across the supply voltage, each resistance produces an I*R voltage drop. We can see that the greatest resistor results in the largest I*R voltage drop using the voltage divider ratio rule. R_1 is therefore 4V and R_2 is 8V.

Kirchhoff's Voltage Law, which states that 4V + 8V = 12V, demonstrates that the total of the voltage drops within the resistive circuit is precisely equal to the supply voltage. The voltage dropped across each resistor would be exactly half the supply voltage for two resistances connected in series if we used two resistors of identical value, meaning R_1 = R_2. This is because the voltage divider ratio would be equal to 50%.

A voltage divider network can also be used to generate an output with a changing voltage. When resistor R_2 is replaced with a variable resistor (potentiometer), the ratio of the two resistive values changes since we have one fixed and one variable resistor, which allows us to adjust the voltage dropped across R_2 and, consequently, V_{OUT}.

Devices with changeable voltage division include potentiometers, trimmers, rheostats, and variacs. By substituting a sensor, such as a light dependent resistor, or LDR, for the fixed resistor R_2, we could extend the concept of variable voltage division.

As a result, the output voltage V_{OUT} fluctuates in direct proportion to how the sensor's resistive value changes in response to variations in light levels. Other types of resistive sensors include thermometers and strain gauges.

The two equations for voltage division discussed above must consequently be related mathematically because they both refer to the same common current. Therefore, the voltage dropped across each particular resistor for any number of individual resistors forming a series network is given as:

Voltage Divider Equation

$$V_{R(x)} = V_S \left(\frac{R_X}{R_T} \right)$$

Here: $V_{R(x)}$ is the voltage drop across the resistor, R_X is the value of the resistor, and R_T is the total resistance of the series network. Due to the proportional relationship between each resistance's R and corresponding voltage drop's V, the voltage divider equation can be used to any number of series resistances coupled together. However, keep in mind that this equation is provided for an unloaded voltage divider network that has no parallel branch currents or extra associated resistive loads.

Example-2 of Voltage Divider

Three resistive components of 6kΩ, 12kΩ and 18kΩ are connected together in series across a 36 volt supply. Calculate, the total resistance, the value of the current flowing around the circuit, and the voltage drops across each resistor.

Data given: V_S = 36 volts, R_1 = 6kΩ, R_2 = 12kΩ and R_3 = 18kΩ

$$R_T = R_1 + R_2 + R_3 = 6k\Omega + 12k\Omega + 18k\Omega = 36k\Omega$$

$$I = \frac{V_S}{R_T} = \frac{36}{36000} = 1mA$$

$$V_{R1} = V_S \left(\frac{R_1}{R_T}\right) = 36\left(\frac{6000}{36000}\right) = 6 \text{volts}$$

$$V_{R2} = V_S \left(\frac{R_2}{R_T}\right) = 36\left(\frac{12000}{36000}\right) = 12 \text{volts}$$

$$V_{R3} = V_S \left(\frac{R_3}{R_T}\right) = 36\left(\frac{18000}{36000}\right) = 18 \text{volts}$$

Voltage Divider Circuit

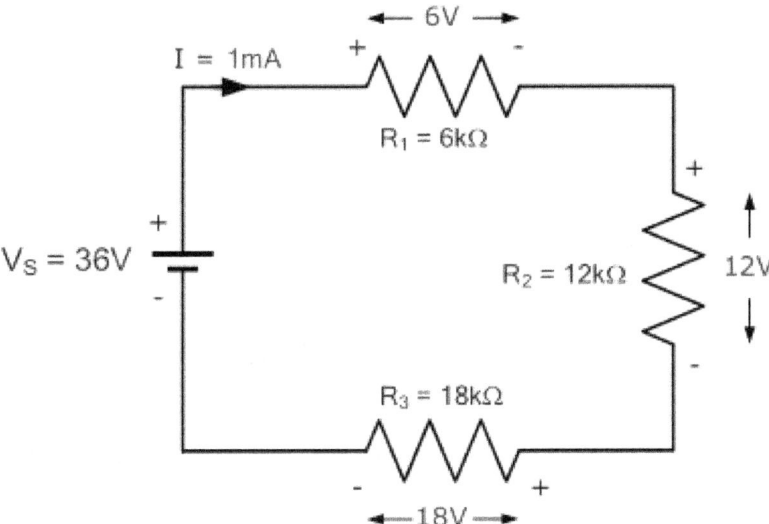

The voltage drops across all three resistors should add up to the supply voltage as defined by Kirchhoff's Voltage Law (KVL). So the sum of the voltage drops is: V_T = 6 V + 12 V + 18 V = 36.0 V the same value of the supply voltage, V_S and so is correct. Again notice that the largest resistor produces the largest voltage drop.

Voltage Tapping Points in a Divider Network

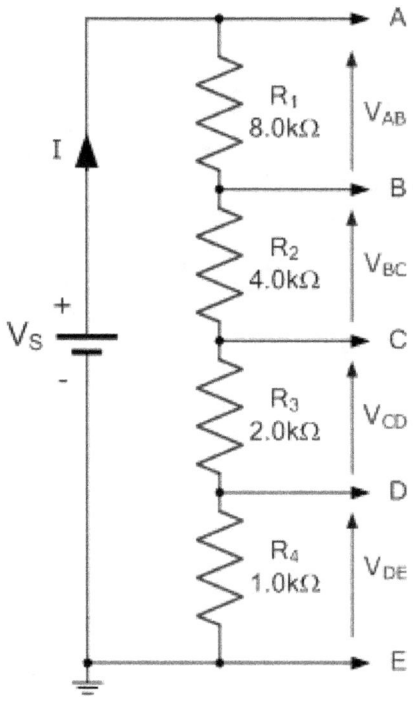

Consider a long series of resistors connected to a voltage source, V_S. Along the series network there are different voltage tapping points, A, B, C, D, and E.

The total series resistance can be found by simply adding together the individual series resistance values giving a total resistance, R_T value of 15kΩ. This resistive value will limit the flow of current through the circuit produced by the supply voltage, V_S.

The individual voltage drops across the resistors are found using the equations above, so $V_{R1} = V_{AB}$, $V_{R2} = V_{BC}$, $V_{R3} = V_{CD}$, and $V_{R4} = V_{DE}$.

The voltage levels at each tapping point is measured with respect to ground (0V). Thus the voltage level at point D will be equal to V_{DE}, and the voltage level at point C will be equal to $V_{CD} + V_{DE}$. In other words, the voltage at point C is the sum of the two voltage drops across R_3 and R_4.

So hopefully we can see that by choosing a suitable set of resistive values, we can produce a sequence of voltage drops which will have a proportional voltage value obtained from a single supply voltage. Note also that in this example each output voltage point will be positive in value because the negative terminal of the voltage supply, V_S is grounded.

Example-3 of Voltage Divider

1. Calculate the no load voltage output for each tapping point of the voltage divider circuit above if the series-connected resistive network is connected to a 15 volt DC supply.

$$R_T = R_1 + R_2 + R_3 + R_4 = 8k\Omega + 4k\Omega + 2k\Omega + 1k\Omega = 15k\Omega$$

$$V_{R1} = V_{AB} = V_S\left(\frac{R_1}{R_T}\right) = 15\left(\frac{8000}{15000}\right) = 8\,\text{volts}$$

$$V_{R2} = V_{BC} = V_S\left(\frac{R_2}{R_T}\right) = 15\left(\frac{4000}{15000}\right) = 4\,\text{volts}$$

$$V_{R3} = V_{CD} = V_S\left(\frac{R_3}{R_T}\right) = 15\left(\frac{2000}{15000}\right) = 2\,\text{volts}$$

$$V_{R4} = V_{DE} = V_S\left(\frac{R_4}{R_T}\right) = 15\left(\frac{1000}{15000}\right) = 1\,\text{volts}$$

2. Calculate the no load voltage output from between points B and E.

$$R_T = R_1 + R_2 + R_3 + R_4 = 8k\Omega + 4k\Omega + 2k\Omega + 1k\Omega = 15k\Omega$$

$$V_{BE} = V_S \left(\frac{R_2 + R_3 + R_4}{R_T} \right) = 15 \left(\frac{4k\Omega + 2k\Omega + 1k\Omega}{15k\Omega} \right) = 7 \text{ volts}$$

A Negative and Positive Voltage Divider

In the simple voltage divider circuit above all the output voltages are referenced from a common zero-voltage ground point, but sometimes it is necessary to produce both positive and negative voltages from a single source voltage supply. For example, the different voltage levels from a computer PSU, -12V, +3.3V, +5V and +12V, with respect to a common reference ground terminal.

Example-4 of Voltage Divider

Applying Ohm's Law, find the values of resistors R_1, R_2, R_3 and R_4 required to produce the voltage levels of -12V, +3.3V, +5V and +12V if the total power supplied to the unloaded voltage divider circuit is 24 volts DC, 60 watts.

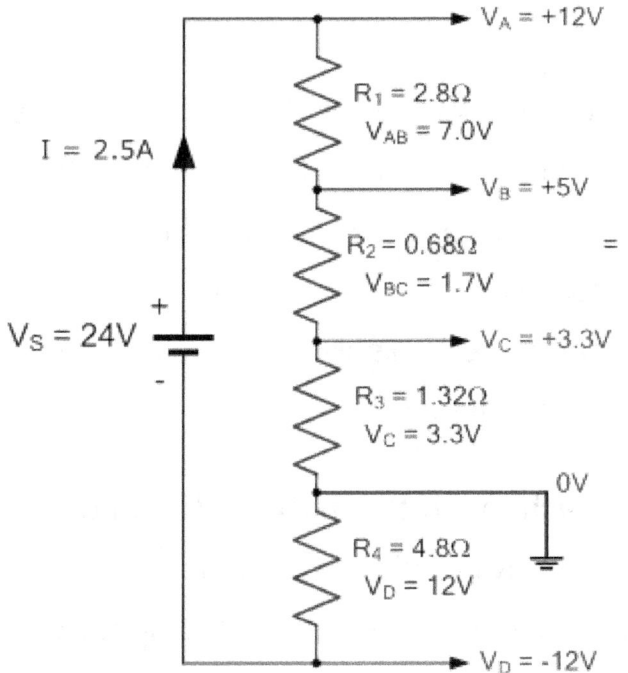

In this example, the zero-voltage ground reference point has been moved to produce the required positive and negative voltages, while maintaining the voltage divider network across the supply. Thus the four voltages are all measured with respect to this common reference point resulting in point D being at the required negative potential of -12V with respect to ground.

We know that series resistive circuits can be used to create a voltage divider, or potential divider network which can be widely used in electronic circuits. By selecting appropriate values for the series resistances, any value of output voltage can be obtained which is lower than the input or supply voltage. But as well as using resistances and a DC supply voltage to create a *resistive voltage divider network*, we can also use capacitors (C) and inductors (L), but with a sinusoidal AC supply as capacitors and inductors are reactive components, meaning that their resistance "reacts" against the flow of electric current.

Capacitive Voltage Dividers

Capacitive Voltage Divider circuits produce voltage drops across capacitors connected in series to a common AC supply. Generally capacitive voltage dividers are used to "step-down" very high voltages to provide a low voltage output signal which can then be used for protection or metering. Nowadays, high frequency capacitive voltage dividers are used more in display devices and touch screen technologies found in mobile phones and tablets.

Unlike resistive voltage divider circuits which operate on both AC and DC supplies, voltage division using capacitors is only possible with a sinusoidal AC supply. This is because the voltage division between series connected capacitors is calculated using the reactance of the capacitors, X_C which is dependent on the frequency of the AC supply.

We know that in AC circuit, capacitive reactance, X_C (measured in Ohms) is inversely proportional to both frequency and capacitance, and is therefore given by the following equation of:

Capacitive Reactance Formula

$$X_C = \frac{1}{2\pi f C}$$

Here:

  X_C = Capacitive Reactance in Ohms, (Ω)

  π (pi) = a numeric constant of 3.142

  f = Frequency in Hertz, (Hz)

C = Capacitance in Farads, (F)

Hence, by knowing the voltage and frequency of the AC supply, we can calculate the reactance of the individual capacitors, substitute them in the above equation for the resistive voltage divider rule, and obtain the corresponding voltage drops across each capacitor as shown.

Capacitive Voltage Divider

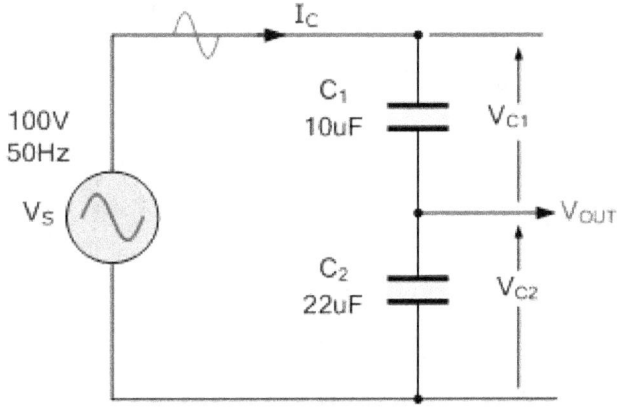

Utilizing the two capacitors of 10uF and 22uF in the series circuit above, we can calculate the rms voltage drops across each capacitor in terms of their reactance when connected to a 100 volts, 50Hz rms supply.

$$X_{C1} = \frac{1}{2\pi f C_1} = \frac{1}{2\pi \times 50 \times 10 \times 10^{-6}} = 318.3\,\Omega$$

$$X_{C2} = \frac{1}{2\pi f C_2} = \frac{1}{2\pi \times 50 \times 22 \times 10^{-6}} = 144.7\,\Omega$$

$$X_{CT} = X_{C1} + X_{C2} = 318.3\,\Omega + 144.7\,\Omega = 463\,\Omega$$

$$V_{C1} = V_S\left(\frac{X_{C1}}{X_{CT}}\right) = 100\left(\frac{318.3}{463}\right) = 69\,\text{volts}$$

$$V_{C2} = V_S\left(\frac{X_{C2}}{X_{CT}}\right) = 100\left(\frac{144.7}{463}\right) = 31\,\text{volts}$$

While utilizing pure capacitors, the sum of all the series voltage drops equals the source voltage, the same as for series resistances. While the amount of voltage drop across each capacitors is proportional to its reactance, it is inversely proportional to its capacitance.

Therefore, the smaller 10uF capacitor has more reactance (318.3Ω) so therefore a greater voltage drop of 69 volts compared to the larger 22uF capacitor which has a reactance of 144.7Ω and a voltage drop of 31 volts respectively. The current in the series circuit, I_C will be 216mA, and is the same value for C_1 and C_2 as they are in series.

After reviewing capacitive voltage divider circuits, it has been observed that as long as there is no series resistance, purely capacitive, the two capacitor voltage drops of 69 and 31 volts will arithmetically be equal to the supply voltage of 100 volts as the two voltages produced by the capacitors are in-phase with each other. If for whatever reason the two voltages are out-of-phase with each other then we cannot just simple add them together as we would apply Kirchhoffs voltage law, but instead phasor addition of the two waveforms would be required.

Inductive Voltage Dividers

Inductive Voltage Dividers produce voltage drops across inductors or coils connected together in series to a common AC supply. An *inductive voltage divider* can consist of a single winding or coil which is divided into two sections where the output voltage is taken from across one of the section, or from two individual coils connected together. The most common example of an inductive voltage divider is the *auto-transformer* with multiple tapping points along its secondary winding.

Inductors act as a short circuit if it is used with steady state DC supplies or with sinusoids having a very low frequency, approaching 0 Hz. As their reactance is almost zero allowing any DC current to easily pass through them, so like the previous capacitive voltage divider network, we must perform any inductive voltage division using a sinusoidal AC supply. Inductive voltage division between series connected inductors can be calculated using the reactance of the inductors, X_L which like *capacitive inductance*, is dependent on the frequency of the AC supply.

We know that inductive reactance, X_L (also measured in Ohms) is proportional to both frequency and inductance so any increases in the supply frequency increases an inductors reactance. Thus *inductive reactance* is defined as:

Inductive Reactance Formula

$$X_L = 2\pi f L$$

Here:

X_L = Inductive Reactance in Ohms, (Ω)

π (pi) = a numeric constant of 3.142

f = Frequency in Hertz, (Hz)

L = Inductance in Henry, (H)

Knowing the voltage and frequency of the AC supply, we can calculate the reactance of the two inductors and use them along with the voltage divider rule to obtain the voltage drops across each inductor as shown.

Inductive Voltage Divider

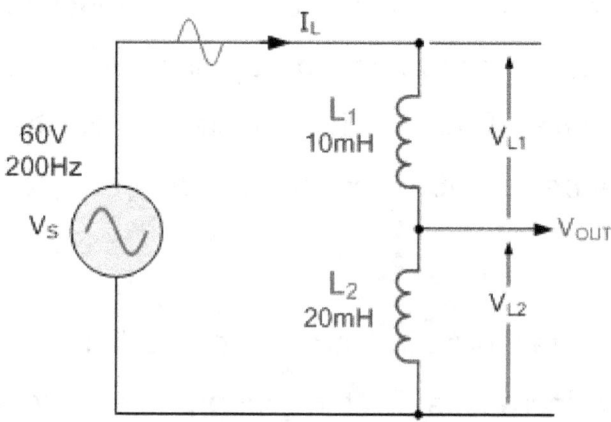

Utilizing the two inductors of 10mH and 20mH in the series circuit above, we can calculate the rms voltage drops across each capacitor in terms of their reactance when connected to a 60 volts, 200Hz rms supply.

$$X_{L1} = 2\pi f L_1 = 2\pi \times 200 \times 10 \times 10^{-3} = 12.56 \Omega$$

$$X_{L2} = 2\pi f L_2 = 2\pi \times 200 \times 20 \times 10^{-3} = 25.14 \Omega$$

$$X_{LT} = X_{L1} + X_{L2} = 12.56 \Omega + 25.14 \Omega = 37.7 \Omega$$

$$V_{L1} = V_S \left(\frac{X_{L1}}{X_{LT}} \right) = 60 \left(\frac{12.56}{37.7} \right) = 20 \, \text{volts}$$

$$V_{L2} = V_S \left(\frac{X_{L2}}{X_{LT}} \right) = 60 \left(\frac{25.14}{37.7} \right) = 40 \, \text{volts}$$

Similar to the previous resistive and capacitive voltage division circuits, the sum of all the series voltage drops across the inductors will equal the source voltage, as long as there are no series resistances. Meaning a pure inductor. The amount of voltage drop across each inductor is proportional to its reactance.

The outcome is that the smaller 10mH inductor has less reactance (12.56Ω), so therefore less of a voltage drop at 30 volts compared to the larger 20mH inductor which has a reactance of 25.14Ω and a voltage drop of 40 volts respectively. The current, I_L in the series circuit is 1.6mA, and will be the same value for L_1 and L_2 as these two inductors are connected in series.

Voltage Divider in Brief

It has been observed that the voltage divider, or network is a very common and useful circuit configuration allowing us to produce different voltage levels from a single voltage supply, thus eliminating the need to have separate power supplies for different parts of a circuit operating at different voltage levels.

A voltage or potential divider, "divides" a fixed voltage into precise proportions using resistors, capacitors or inductors. The most basic and commonly used voltage divider circuit is that of two fixed-value series resistors, but a potentiometer or rheostat can also be used for voltage division by simply adjusting its wiper position.

A voltage divider circuit is usually applied to replace one of the fixed-value resistors with a sensor. Resistive sensors such as light sensors, temperature sensors, pressure sensors and strain gauges, which change their resistive value as they respond to environmental changes can all be used in a voltage divider network to provide an analogue voltage output. The biasing of bipolar transistors and MOSFETs is also another common application of a Voltage Divider.

CHAPTER-16: CURRENT DIVIDERS

Current Divider circuits consists two or more parallel branches for currents to flow through but the voltage is the same for all components in the parallel circuit

Current Divider Circuits are parallel circuits in which the source or supply current divides into a number of parallel paths. In a parallel connected circuit, all the components have their terminals connected together sharing the same two end nodes. This results in different paths and branches for the current to flow or pass along. However, the currents can have different values through each component.

The core characteristic of parallel circuits is that while they may produce different currents flowing through different branches, the voltage is common to all the connected paths. That is $V_{R1} = V_{R2} = V_{R3}$ ... etc. Hence, the need to find the individual resistor voltages is eliminated allowing branch currents to be easily found with Kirchhoff's Current Law, (KCL) and of course Ohm's Law.

Resistive Voltage Divider

The easiest to understand, and most basic form of a passive current divider network is that of two resistors connected together in parallel. The *Current Divider Rule* helps us to calculate the current flowing through each parallel resistive branch as a percentage of the total current. Consider the circuit below.

Resistive Current Divider Circuit

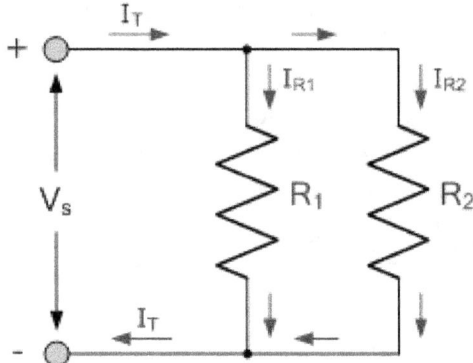

In this circuit, basic current divider circuit consists of two resistors: R_1, and R_2 in parallel which splits the supply or source current I_S between them into two separate currents I_{R1} and I_{R2} before joining together again and returning back to the source.

As the source or total current equals the sum of the individual branch currents, then the total current, I_T flowing in the circuit is given by Kirchoffs current law KCL as being:

$$I_T = I_{R1} + I_{R2}$$

Since the two resistors are connected in parallel, for Kirchhoff's Current Law, (KCL) to hold true it must therefore follow that the current flowing through resistor R_1 will be equal to:

$$I_{R1} = I_T - I_{R2}$$

and the current flowing through resistor R_2 will be equal to:

$$I_{R2} = I_T - I_{R1}$$

As the same voltage, (V) is present across each resistive element, we can find the current flowing through each resistor in terms of this common voltage as it is simply V = I*R following Ohm's Law. So solving for the voltage (V) across the parallel combination gives us:

$$I_T = I_{R1} + I_{R2}$$

$$I_{R1} = \frac{V}{R_1} \quad \text{and} \quad I_{R2} = \frac{V}{R_2}$$

$$I_T = \frac{V}{R_1} + \frac{V}{R_2} = V\left[\frac{1}{R_1} + \frac{1}{R_2}\right]$$

$$\therefore V = I_T\left[\frac{1}{R_1} + \frac{1}{R_2}\right]^{-1} = I_T\left[\frac{R_1 R_2}{R_1 + R_2}\right]$$

Solving for I_{R1} gives:

$$I_{R1} = \frac{V}{R_1} = I_T\left[\frac{\frac{1}{R_1}}{\frac{1}{R_1} + \frac{1}{R_2}}\right]$$

$$\therefore I_{R1} = I_T\left(\frac{R_2}{R_1 + R_2}\right)$$

Similarly, solving for I$_{R2}$ gives:

$$I_{R2} = \frac{V}{R_2} = I_T \left[\frac{\frac{1}{R_2}}{\frac{1}{R_1} + \frac{1}{R_2}} \right]$$

$$\therefore I_{R2} = I_T \left(\frac{R_1}{R_1 + R_2} \right)$$

Have a look at the above equations. Here, each branch current has the opposite resistor in its numerator. That is to solve for I$_1$ we use R$_2$, and to solve for I$_2$ we use R$_1$. This is because each branch current is inversely proportional to its resistance resulting in the smaller resistance having the larger current.

Example-1 of Current Divider

A 20Ω resistor is connected in parallel with a 60Ω resistor. If the combination is connected across a 30 volts battery supply, find the current flowing through each resistor and the total current supplied by the source.

$$I_{R1} = \frac{V}{R_1} = \frac{30}{20} = 1.5 \text{ Amps}$$

$$I_{R2} = \frac{V}{R_2} = \frac{30}{60} = 0.5 \text{ Amps}$$

$$I_T = I_{R1} + I_{R2} = 1.5 + 0.5 = 2.0 \text{ Amperes}$$

Keep in mind that the smaller 20Ω resistor has the larger current because by its very nature, a greater current will always flow through the path or branch of least resistance. This implies then that a short-circuit will produce maximum current flow, while an open-circuit will result in zero current flow. Remember also that the equivalent resistance, R_{EQ} of parallel connected resistors will always be less than the ohmic value of the smallest resistor with the equivalent resistance decreasing as more parallel resistances are added.

In some cases, it is not necessary to calculate all the branch currents, if the supply or total current, I_T is already known, then the final branch current can be found by simply subtracting the calculated currents from the total current as defined by Kirchhoffs current law.

Example-2 of Current Divider

Three resistors are connected together to form a current divider circuit as shown below. If the circuit is fed from a 100 volts 1.5kW power supply, calculate the individual branch currents using the current division rule and the equivalent circuit resistance.

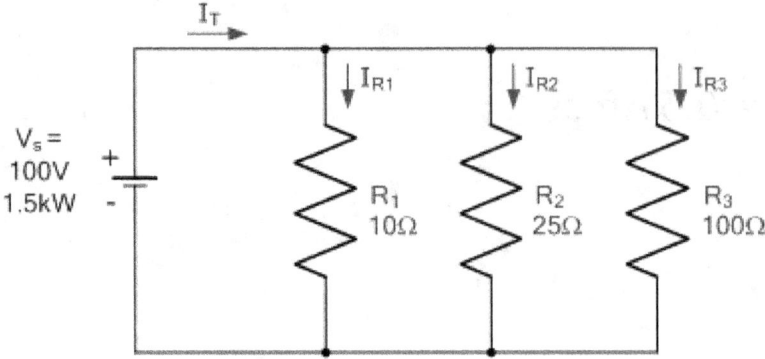

1) Total circuit current I_T

$$P = V_S \times I_T$$

$$I_T = \frac{P}{V} = \frac{1500}{100} = 15\,\text{Amps}$$

2) Equivalent resistance R_{EQ}

$$R_{EQ} = \left[\cfrac{1}{\cfrac{1}{R_1} + \cfrac{1}{R_2} + \cfrac{1}{R_3}} \right]$$

$$R_{EQ} = \left[\cfrac{1}{\cfrac{1}{10} + \cfrac{1}{25} + \cfrac{1}{100}} \right]$$

$$R_{EQ} = \frac{1}{0.15} = 6.667 \, \Omega$$

3) Branch currents I_{R1}, I_{R2}, I_{R3}

$$I_{R1} = I_T \left(\frac{R_{EQ}}{R_1} \right) = 15 \left(\frac{6.667}{10} \right) = 10 \text{ Amps}$$

$$I_{R2} = I_T \left(\frac{R_{EQ}}{R_2} \right) = 15 \left(\frac{6.667}{25} \right) = 4 \text{ Amps}$$

$$I_{R3} = I_T \left(\frac{R_{EQ}}{R_3} \right) = 15 \left(\frac{6.667}{100} \right) = 1 \text{ Amps}$$

According to Kirchhoff's Current Rule, we can check our calculations. Here, all the branch currents will be equal to the total current, so: $I_T = I_{R1} + I_{R2} + I_{R3}$ = 10 + 4 + 1 = 15 amperes, as expected. Therefore, we can see that the total current, I_T is divided according to a simple ratio determined by the branch resistances. Also, as the number of resistors connected in parallel increases, the supply of total current, I_T will also increase for a given supply voltage, V_S as there are more parallel branches taking current.

Current Division utilizing Conductance

Another modest way to find the branch currents in a parallel circuit is to use the conductance method. In DC circuits, Conductance is the reciprocal of resistance, and is denoted by the letter "G". As conductance (G) is the reciprocal of resistance (R) which is measured in Ohm's (Ω), the reciprocal of Ohm's is called "mho" (℧), (an inverted ohm sign). Thus G = 1/R. The electrical units given to conductance is the Siemen (symbol S).

Therefore, for parallel connected resistors, the equivalent or total conductance, C_T will be equal to the sum of the individual conductance as shown.

Parallel Conductance

$$\frac{1}{R_T} = \frac{1}{R_1} + \frac{1}{R_2} + \frac{1}{R_3} + \ldots etc$$

$$G_T = G_1 + G_2 + G_3 + \ldots etc$$

Hence, if a resistance has a fixed value of 10Ω, it will have an equivalent conductance of 0.1S and so on. Because of the reciprocal, a high value of conductance represents a low value of resistance, and vice versa. We can also use prefixes in the form of *milli-Siemens*, mS, *micro-Siemens*, uS and even *nano-Siemens*, nS for very small conductance. So a resistor of 10kΩ will have a conductance of 100uS.

Applying the Ohm's Law equation for current in which I = V/R, we can define the branch currents using conductance as being: I = V*G

Indeed, we can take this one step further by saying that the supply current to our parallel resistive network above is:

$$I_S = G_T \times V_S$$

$$\therefore I_S = (G_1 + G_2 + G_3) \times V$$

From the above, we understand that for a parallel connected circuit, voltage is common to all components and as voltage equals current times resistance, V = I*R, we can therefore conclude that when using conductance, the voltage is equal to current divided by conductance. That is V = I/G.

Then we can express the above equations for the current divider rule in relationship to conductance (G), instead of the resistance (R) as being:

Current Divider Rule utilizing Conductance

$$I_{R1} = G_1 \times V = G_1\left(\frac{I_T}{G_T}\right)$$

$$\therefore I_{R1} = I_T\left(\frac{G_1}{G_T}\right)$$

Similarly, for the currents in parallel resistors R_2 and R_3 are given as:

$$I_{R2} = I_T\left(\frac{G_2}{G_T}\right); \quad I_{R3} = I_T\left(\frac{G_3}{G_T}\right)$$

It has been observed that unlike the equations above for resistance, each branch current has the same conductance in its numerator. That is to solve for I_1 we use G_1, and to solve for I_2 we use G_2. This is because the conductance are the reciprocals of the resistances.

Example-3 of Current Divider

Applying the conductance method, determine the individual branch currents, I_1, I_2 and I_3 of the following parallel resistive circuit.

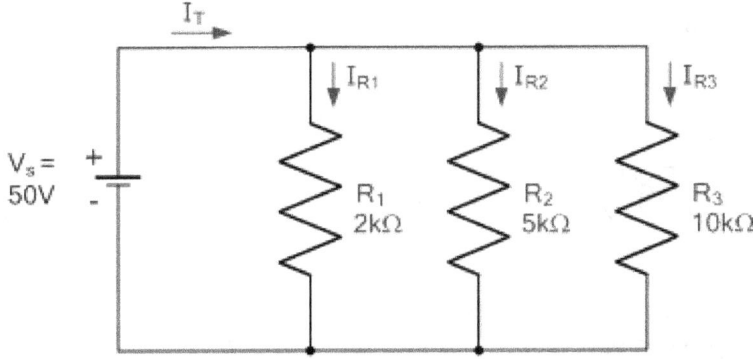

Total conductance G_T

$$G_T = \frac{1}{R_1} + \frac{1}{R_2} + \frac{1}{R_3} = \frac{1}{2000} + \frac{1}{5000} + \frac{1}{10000}$$

$$\therefore G_T = \frac{8}{10000} = \frac{1}{1250} = 800\,\mu S$$

Total supply current I_S

$$I_T = V_S \times G_T = 50 \times 0.0008 = 0.04\,A \text{ or } 40\,mA$$

$$G_1 = \frac{1}{2000} = 500\,\mu S$$

$$G_2 = \frac{1}{5000} = 200\,\mu S$$

$$G_3 = \frac{1}{10000} = 100\,\mu S$$

Individual branch currents I_1, I_2 and I_3

$$I_{R1} = I_T\left(\frac{G_1}{G_T}\right) = 0.04\left(\frac{0.0005}{0.0008}\right) = 25\,\text{mA}$$

$$I_{R2} = I_T\left(\frac{G_2}{G_T}\right) = 0.04\left(\frac{0.0002}{0.0008}\right) = 10\,\text{mA}$$

$$I_{R3} = I_T\left(\frac{G_3}{G_T}\right) = 0.04\left(\frac{0.0001}{0.0008}\right) = 5\,\text{mA}$$

Since conductance is the reciprocal or inverse of resistance, the equivalent resistance value of the example circuit is simply 1/800uS which equals 1250Ω or 1.25kΩ, which is clearly less than the smallest resistor value of R_1 at 2kΩ.

Current Divider in A Brief

The process of determining the individual branch currents in a parallel circuit wherein each parallel element has the same voltage is known as Current dividers or current division. Kirchhoff's current law, (KCL) states that the algebraic sum of the individual currents entering a junction or node will equal the currents leaving it. That is the net result is zero.

In order to determine individual branch currents when the equivalent resistance and the total circuit current are known, Kirchhoff's current divider rule can also be used. When only two resistive branches are involved, the current in one branch will be some fraction of the total current I_T. If the two parallel resistive branches are of equal value, the current will divide equally.

For three or more parallel branches, the equivalent resistance R_{EQ} is used to divide the total current into the fractional currents for each branch producing a current ratio which is equal to the inverse of their resistive values resulting in the smaller value resistance having the greatest share of the current. The supply or total current, I_T being the sum of all the individual branch currents. This then makes current dividers useful for use with current sources.

It is often convenient to use conductance with parallel circuits as it can help reduce the math required for determining the branch currents through individual circuit elements that are connected together in parallel. This is because for parallel circuits the total conductance is the sum of the individual conductance values. Conductance is the reciprocal or inverse of resistance as $G = 1/R$. The units for conductance are Siemens, S. The conductance of an element can also be used even if the supply voltage is DC or AC for *current dividers*.

CHAPTER-17: ELECTRICAL ENERGY AND POWER

Electrical Energy supplies the power required to produce work or an action within an electrical circuit and is given in joules per second

Electrical Energy is the capability of an electrical circuit to produce work by creating an action. This action can take many forms, such as thermal, electromagnetic, mechanical, electrical, etc. *Electrical energy* can be both created from batteries, generators, dynamos, and photovoltaics, etc. or stored for future use using fuel cells, batteries, capacitors or magnetic fields, etc. Thus electrical energy can be either created or stored.

Energy cannot be created or destroyed, only converted. But for energy to do any useful work it must be converted from one form into something else. For example, a motor converts electrical energy into mechanical or kinetic (rotational) energy, while a generator converts kinetic energy back into electrical energy to power a circuit.

That is electrical machines convert or change energy from one form to another by doing work. Another example is a lamp, light bulb or LED (light emitting diode) which convert electrical energy into light energy and heat (thermal) energy. Then electrical energy is very versatile as it can be easily converted into many other different forms of energy.

For electrical energy to move electrons and produce a flow of current around a circuit, work must be done, that is the electrons must move by some distance through a wire or conductor. The work done is stored in the flow of electrons as energy. Thus "Work" is the name we give to the process of energy.

It has been observed that *Work* and *Energy* are effectively the same as energy can be defined as "the ability to do some work". Keep in mind that work done or energy transferred applies equally to a mechanical system or thermal system as it does to an electrical system. Since mechanical, thermal and electrical energies are interchangeable, this is happened.

Electrical Energy: The Volt

The capacity to do work is called energy. The standard unit used for energy (and work) is the Joule. A joule of energy is defined as the energy expended by one ampere at one volt, moving in one second. Electric current results from the movement of electric charge (electrons) around a circuit, but to move charge from one node to another there needs to be a force to create the work to move the charge, and there is: *voltage*.

We imagine voltage (V) as existing between two different terminals, points or nodes within a circuit or battery supply. But voltage is important as it provides the work needed to move the charge from one point to another, either in a forward direction or a reverse direction. The voltage, or potential difference between two terminals or points is defined as having a value of one volt, when one joule of energy is used in moving one coulomb of electric charge between those two terminals.

In other words, the *Voltage* difference between two points or terminals is the work required in *Joules* to move one *Coulomb* of charge from A to B. Hence, voltage can be expressed as being:

The Voltage Unit

$$1\,\text{Volt} = \frac{1\,\text{Joule}}{1\,\text{Coulomb}} = \frac{J}{C}$$

Here: voltage is in Volts, J is the work or energy in Joules and C is the charge in Coulombs. Thus if J = 1 joule, C = 1 coulomb, then V will equal 1 volt.

Example-1 of Electrical Energy

What is the terminal voltage of a battery which requires 135 joules of energy to move 15 coulombs of charge around an electrical circuit?

$$\text{voltage} = \frac{\text{energy}}{\text{charge}} = \frac{J}{C} = \frac{135}{15} = 9V$$

In this example, we observed that every coulomb of charge possesses an energy of 9 joules.

Electrical Energy: The Ampere

It has been observed that the unit of electrical charge is the *Coulomb* and that the flow of electrical charge around a circuit is used to represent a flow of current. The common symbol used for electrical charge is the capital letter "Q" or small letter "q", basically standing for quantity. Thus Q = 1 coulomb of charge or Q = 1C. Note that charge Q can be either positive, +Q or negative, -Q, that is an excess of either electrons or holes.

Charge flow around a closed circuit in the form of electrons is called an *electric current*. However, the use of the expression "charge flow" denotes movement, so to produce an electrical current, charge must move. This then leads to the question of what is making the charge move, and this is done by our old friend Voltage from above.

Thus the voltage or potential difference between two points provides the required electrical energy to move charge around a circuit in the form of an electric current. So, the work done to move charge is provided by a potential difference, and if there is no potential difference between two points, there is no movement of charge and therefore no current flow. Charge without any flow or movement is called static electricity indeed.

When electric charge flows through a specific point of the circuit in exactly one second, this will provide us the strength of the electrical current in *Amperes*, (A). In this way, one ampere of current is equal to one coulomb of charge which flows through a specific point in one unit second, and the more charge per second which passes this point, the greater will be the current. Now, we can define one ampere (A) of electrical current as being equal to one coulomb of charge per second. Therefore, 1A = 1C/s

The Ampere Unit

$$1 \text{ Ampere} = \frac{1 \text{ Coulomb}}{1 \text{ Second}} = \frac{1C}{1s} = \frac{Q}{t}$$

Here: Q is the charge (in coulombs) and t is the interval in time (in seconds) that the charge flows. That is, electrical current has both a magnitude (the amount of charge) and a specified direction associated with it.

Keep in mind that the commonly used symbol for electrical current is the capital letter "I", or small "i" both standing for intensity. That is the intensity or concentration of charge producing the electron flow. For a constant DC current, the capital letter "I" is generally used, whereas for a time-varying AC current, the lower case letter "i" is commonly used. The symbol $i_{(t)}$ simply means an instantaneous current value at that exact instant in time.

It is often easier to remember this relationship by using an image. Here the three quantities of Q, I and t have been superimposed into a triangle represents the actual position of each quantity within the current formula.

The Ampere

Rearranging the standard formula above provides us the following combinations of the same equation:

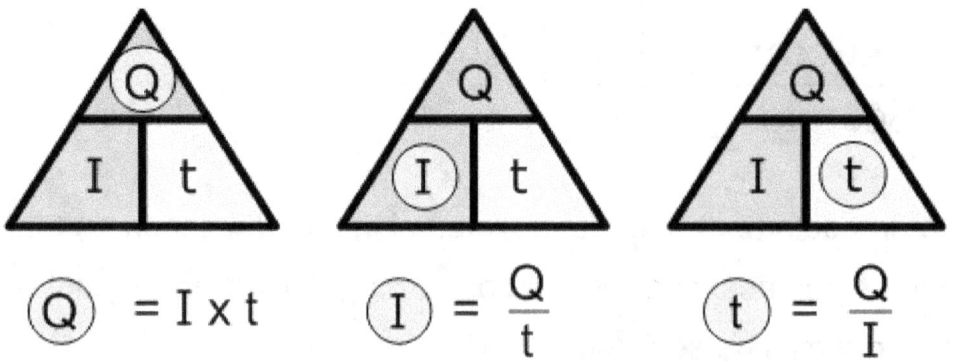

Example-2 of Electrical Energy

1. Find out the current flows through a circuit if 900 coulombs of charge flows through a specific point in 3 minutes.

$$I = \frac{Q}{t} = \frac{900}{3 \times 60} = \frac{900}{180} = 5 \text{ Amperes}$$

2. An electric current of 3 Amperes flows through a resistor. Find out coulombs of charge will flow through the resistor in 90 seconds.

$$Q = I \times t = 3A \times 90s = 270 \text{ Coulombs}$$

Unit of the Electrical Power: The Watt

The rate at which a work is done by utilizing required energy is called Electrical Power. It is the product of the two quantities, Voltage and Current. Voltage provides the work required in Joules to move one Coulomb of charge from A to B and that current is the rate of movement (or rate of flow) of the charge. So how are these two definitions linked together.

When voltage, (V) equals Joules per Coulombs (V = J/C) and Amperes (I) equals charge (*coulombs*) per second (A = Q/t), then we can define electrical power (P) as being the totality of these two quantities. Since electrical power is also equal to voltage times amperes, hence: P = V*I.

The Watt

$$V = \frac{J}{C} \quad \text{and} \quad I = \frac{Q}{t}$$

As: $Q = 1C$

If: $P = V \times I = \frac{J}{C} \times \frac{Q}{t} = \frac{J}{C} \times \frac{C}{t}$

Then: $P = \frac{J}{\cancel{C}} \times \frac{\cancel{C}}{t} = \frac{J}{t}$

Therefore, electrical power is also the rate at which work is done during one second. That is, one joule of energy dissipated in one second. Since electrical power is measured in Watts (W), hence it must be also being measured in *Joules per Second*. Therefore, we can say that: 1 watt = 1 joule per second (J/s).

Electrical Power

1 watt (W) = 1 joule/second (J/s)

When 1 watt = 1 joule per second, it follows that: 1 Joule of energy = 1 watt over one unit of time, that is: Work equals Power multiplied by Time, (V*I*t joules). Therefore, electrical energy (the work done) is obtained by multiplying power by the time in seconds that the charge (in the form of a current) flows.

In this way. units of electrical energy depend on the units used for electric power and time. So if we measure electrical power in kilowatts (kW), and the time in hours (h), then the electrical energy consumed equals kilowatts*hours (Wh) or simply: kilowatt-hours (kWh).

Example-3 of Electrical Energy

A 100 Watt light bulb is illuminated on for one hour only. Find out joules of electrical energy have been used by the lamp.

$$\text{Electrical Energy} = \text{Power} \times \text{Time}$$

$$\text{Electrical Energy} = 100 \times (60 \times 60)$$

$$\text{Electrical Energy} = 100 \times 3600 = 360{,}000 \text{ joules}$$

Keep in mind that when dealing with the joule as a unit of electrical energy, it is more convenient to present them in kilo-joules. So, the answer can be given as: 360kJ.

Since a *joule* on its own is a small quantity, the kilojoule (kJ), thousands of joules, the megajoule (MJ), millions of joules, and even the gigajoule (GJ), thousands of millions of joules, are all practical units of electrical energy. Thus one unit of electricity which is equivalent to one kilowatt-hour (kWh) can be defined as 3.6 megajoules (MJ).

Similarly, as a Watt is a small amount of electrical power, kilowatts (1 kW = 1,000 watts) and megawatts (1 MW = 1 million watts) are usually used to find the power output of electrical equipment and appliances. Thus we can see that the kilowatt (or megawatt) is a unit of electrical power, while the kilowatt-hour is a unit of electrical energy.

CHAPTER-18: DC CIRCUIT AND WAVEFORM

Direct Current (D.C) is a form of electrical current which flows around an electrical circuit in one direction only, making it a "Uni-directional" supply.

Usually, both DC currents and voltages are produced by power supplies, batteries, dynamos and solar cells to name a few. A DC voltage or current has a fixed magnitude (amplitude) and a definite direction associated with it. For example, +12V represents 12 volts in the positive direction, or -5V represents 5 volts in the negative direction.

DC power supplies do not change their value with regards to time, they are a constant value flowing in a continuous steady state direction. That is, DC maintains the same value for all times and a constant uni-directional DC supply never changes or becomes negative unless its connections are physically reversed. An example of a simple DC or direct current circuit is given below.

DC Circuit and Waveform

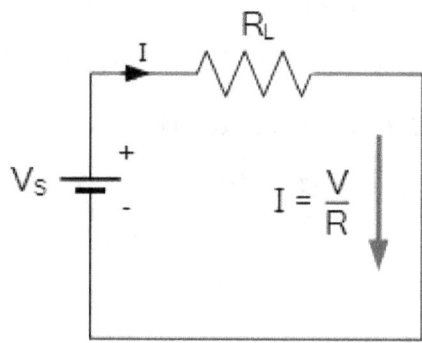 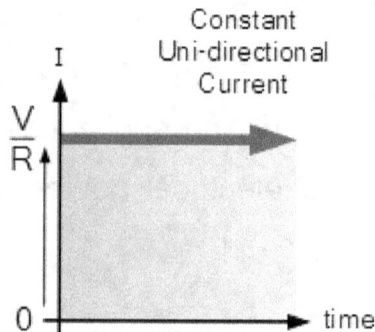

www.ingramcontent.com/pod-product-compliance
Lightning Source LLC
Chambersburg PA
CBHW060416220526
45465CB00008B/2906